Oscarmetrics

Ben Zauzmer

Oscarmetrics
© 2019. Ben Zauzmer. All rights reserved.

No part of this book may be reproduced in any form or by any means, electronic, mechanical, digital, photocopying or recording, except for the inclusion in a review, without permission in writing from the publisher.

Published in the USA by:
BearManor Media
P O Box 71426
Albany, Georgia 31708
www.bearmanormedia.com

Printed in the United States of America
ISBN   978-1-62933-440-0 (paperback)
            78-1-62933-441-7 (hardback)

Book and cover design by Darlene Swanson • www.van-garde.com

DISCLAIMER: "OSCAR®," "OSCARS®," "ACADEMY AWARD®," "ACADEMY AWARDS®," "OSCAR NIGHT®," "A.M.P.A.S.®" and the "Oscar" design mark are trademarks and service marks of the Academy of Motion Picture Arts and Sciences. The appearance of the registered trademark symbol ® is implied wherever the terms Oscar, Academy Award, or Oscar Night, or the plural versions of these terms, appear in this book. This book is neither authorized nor endorsed by the Academy of Motion Pictures Arts and Sciences.

Contents

Opening Monologue vii
Chapter 1.   Best Film Editing. 1
Chapter 2.   Best Visual Effects 7
Chapter 3.   Best Supporting Actor 15
Chapter 4.   Best Supporting Actress. 23
Chapter 5.   Best Makeup and Hairstyling. 31
Chapter 6.   Best Costume Design. 41
Chapter 7.   Best Production Design. 51
Chapter 8.   Best Cinematography. 65
Chapter 9.   Best Original Score 77
Chapter 10. Best Original Song 87
Chapter 11. Best Sound Editing. 101
Chapter 12. Best Sound Mixing 109
Chapter 13. Short Film Categories 115
Chapter 14. Best Animated Feature 121
Chapter 15. Best Documentary 129
Chapter 16. Best Foreign Language Film 137
Chapter 17. Best Popular Film. 145
Chapter 18. Best Original Screenplay 155
Chapter 19. Best Adapted Screenplay 161

Chapter 20.	Best Actor	169
Chapter 21.	Best Actress	177
Chapter 22.	Best Director	187
Chapter 23.	Best Picture	197
Chapter 24.	After-Party	207
Epilogue.	Predicting the Oscars	247
	Acknowledgments	253
	Index	255

I'd like to thank the Academy.

Opening Monologue

MARIA:

Let's start at the very beginning. A very good place to start.

—*The Sound of Music* (1965): Nominated for 10 Oscars; won 5 (Best Picture, Best Director, Best Musical Score, Best Film Editing, Best Sound)

LAUREN BACALL, the legend of stage and screen, was finally on the verge of receiving her first Oscar. After a career of starring in movies from *To Have and Have Not* (1944) to John Wayne's final film, *The Shootist* (1976), this was her moment. She had won a Screen Actors Guild award and a Golden Globe for *The Mirror Has Two Faces* (1996), and now many viewed the subsequent Oscar as a mere formality.

Juliette Binoche was the relative newcomer. Though she had been popular in France for several years, her name recognition in America was a fraction of Bacall's heading into 1996. She played a supporting role in *The English Patient*, which won Best Picture later that evening, and she gained some pre-Oscars momentum with a British Academy of Film and Television Arts (BAFTA) victory that some Oscar predictors ignored.

The envelope opened, the winner was Binoche. A look of shock, a walk to the stage. First words into the microphone: "I'm so surprised. It's true I didn't prepare anything. I thought Lauren was going to get it."

Why? Did the sentimentality of Bacall's topping off her lustrous career with an Oscar after more than half a century seem insurmountable? Does a Screen Actors Guild win plus a Golden Globe guarantee an Oscar? In truth, there were plenty of signs to indicate a close race, but neither Binoche nor the media noticed.

If you've ever read articles by expert Oscar columnists in the run-up to awards season, they're likely relying on traditional reporting techniques like interviews and film analysis, not data and calculators.

That's where I come in.

Rewind a few years to my freshman year at Harvard. With the 2012 presidential election just ten months away, and pitchers and catchers due to report to spring training in a few weeks, two of my favorite mathematical prediction seasons were just around the corner. I studied applied math in college, and what could be a more engaging application of math than trying to predict the future, especially the future of something as entertaining as politics or baseball?

That's when it hit me. What about my other passion: movies? Surely, in the vast realm of the internet, there was a person who had revolutionized the art of Oscar prediction with data, the way Bill James did for baseball and Nate Silver did for politics. Heck, even one of the Oscar nominees that year – *Moneyball* (2011) – focused on mathematically predicting a field traditionally dominated by non-mathematical thinkers. I went to Google and found nothing.

My immediate reaction was that it must be impossible, or someone else would have already stepped in. But I had to know. So, like any good applied math major (or "concentrator" at Harvard, for some unknown reason), I went hunting for a dataset filled with everything

I would need: Oscar results from every year in every category, with all of the pre-Oscars data points for each movie, and all of the same data for the upcoming year's nominees. Again, Google failed me.

Therefore, being the cool college freshman I was, I proceeded to spend a solid month on the third floor of Lamont Library, not only enjoying one of the warmest rooms and comfiest chairs in Cambridge during a typically bitter Massachusetts winter, but also taking advantage of the quiet and Wi-Fi to build my own Oscar dataset. As far as I know, it was the only one quite like it in the world. I used the Academy website, Wikipedia, Rotten Tomatoes, Metacritic, IMDb, and a host of other sources, often even hunting through old press releases to collect individual data points, one at a time.

One month later, dataset finally in hand, I built some formulas. The gist of these formulas lies in weighting data points such as which categories a film is nominated in and which pre-Oscars awards a film has won. More weight goes to the inputs that have done the best job of predicting each category in previous years.

If all of the data points happen to point in the same direction, math is unnecessary, at least as far as determining a favorite goes. But what do we do when the indicators don't agree? Say the BAFTAs pick one nominee for Best Actress, but the Screen Actors Guild picks another. If we're really unlucky, the Golden Globes choose a third and fourth, for their comedy/musical and drama actress categories.

As a matter of fact, that's exactly what happened in 2001. Halle Berry won the Screen Actors Guild award for *Monster's Ball*, and Judi Dench took the BAFTA for *Iris*, while the Golden Globes went to Nicole Kidman for *Moulin Rouge!* and Sissy Spacek for *In the Bedroom*.

Traditionally, we trust human intuition to weight all of these factors appropriately. But to be frank, this is where math can outshine people. There are absolutely some components of Oscar prediction that human intuition handles better than math, but weighting data points isn't one of them.

Once my formulas were ready, I made predictions in 20 Oscar categories and put them up on a decidedly second-rate website I built in a few minutes. The word "Harvard" often garners press coverage for stories that otherwise would never make the news. Same thing goes for the word "Oscars." Turns out, if you put them together, they're powerful enough to make even my basic website the subject of articles from around the world. The pressure was on.

The night of the ceremony arrived, and I co-hosted an Oscar party in one of Harvard's Hogwartsian common rooms while nervously anticipating the results. At first, the evening only went so-so. I started off 9-for-14, much better than flipping a five-sided coin each time (one side for each nominee in the category), but not quite groundbreaking either. Things began to pick up steam as the major categories came in. My picks for both screenplay awards were correct, and Best Director and Best Actor also went according to Oscarmetrics. And then, it was time for Best Actress.

Best Actress that year was, by all accounts, a two-horse race: Viola Davis for *The Help* (2011) versus perennial nominee Meryl Streep for *The Iron Lady* (2011). A consensus was forming around Davis, including by many of journalism's most prominent Oscar predictors, for a variety of reasons – some mathematically justifiable, some not so much. But the math slightly favored Streep by just a fraction of a percentage point.

This was my biggest chance to prove myself, and more importantly, to show that math actually could predict the Oscars. This was the category that could differentiate my predictions from those of the nonmathematical predictors.

Colin Firth, on presenter duty as the most recent Best Actor for *The King's Speech* (2010), opened the envelope. The winner: Meryl Streep. A bridesmaid for twelve unsuccessful nominations in a row, she was once again a bride, and I'm quite sure she was the happiest person in the world. But I may have been second.

Before we jump in, a few ground rules:

- This book contains data through the Oscars held in March 2018. This is not a book about the upcoming Oscars, so if you're looking for a list of picks to win your pool this year, I recommend going to the internet instead. Feel free to Google my name: I write articles every year on just that subject.

- This is not a math textbook. But I will provide enough of an explanation along the way so that anyone who is interested in learning more will know what terms to search. If you haven't the slightest idea what a logistic regression is or what a Bayesian prior means, rest assured, this book is still meant for you. Anyone who is already fluent in statistics will understand what I'm up to and may find some portions to be self-evident.

- On the flip side, while I hope you like movies if you're reading this book, I'm not going to assume any Oscar knowledge. Can you name the three films that won Best Picture, Best Director, Best Actor, Best Actress, and a screenplay honor? Once you've made your guess, you can find the answer in the footnote.[1] If you got it right, that's great. You're going to love this stuff. But it's certainly not required knowledge for understanding anything on these pages.

- Each chapter will explore one specific question about the Academy Awards, with one question per Oscar category, though sometimes multiple categories will make an appearance in a single chapter. I will follow a typical order of categories announced at the ceremony (though the exact order changes annually), starting with technical categories and supporting actor/actress, and eventually concluding with Best Picture.

That's it! Enough of the monologue. The ceremony's about to begin.

---

[1] *It Happened One Night* (1934), a delightful romantic comedy about a reporter and an heiress on the run, won all five categories it was nominated for. Frank Capra directed, while Clark Gable and Claudette Colbert took home acting honors.

One Flew Over the Cuckoo's Nest (1975) portrays a frightening mental institution, in which Nurse Ratched (Louise Fletcher) passive-aggressively controls Randle McMurphy (Jack Nicholson) and the other patients. Milos Forman directed the pair to victory, the first time since 1934 that a film took home both lead acting trophies.

The Silence of the Lambs (1991) tells the creepy story of an FBI agent (Jodie Foster) working with a serial killer (Anthony Hopkins) to hunt down a different murderer. Jonathan Demme directed the only horror film to win Best Picture.

Chapter 1. Best Film Editing

Which category is the best predictor of Best Picture?

> **HERRON:**
> Wait'll you see it. I don't know whether to edit or leave it raw like this.
>
> –*Network* (1976): Nominated for 10 Oscars; won 4 (Best Actor, Best Actress, Best Supporting Actress, Best Original Screenplay)

W₁ BEGIN with a category that is often buried near the front of the ceremony: Best Film Editing. Hardly anyone yells at the crowd during an Oscar viewing party, "Everyone be quiet! Best Editing is next!" More likely, half the guests haven't even arrived yet, and a couple of pizza boxes are still unopened.

That said, for those who pay attention, Best Editing can be a key predictor of the most celebrated award of them all, Best Picture. Since 1981, the only film to win Best Picture without an editing nomination is *Birdman* (2014). And *Birdman* is a unique case, because the whole gimmick is that it appears to be mostly comprised of one uninterrupted, dizzying tracking shot – so clearly, an editing nomination was never in the cards for a film without editing.

Does that mean that Best Editing is the category most closely correlated with Best Picture? Hold on a second: What about Best

Director? Any Oscar pundit would tell you that in order to win Best Picture, a nominated film needs a Best Director nomination as well. Indeed, only four films have proved to be exceptions: *Wings* (1927), *Grand Hotel* (1932), *Driving Miss Daisy* (1989), and *Argo* (2012).

And what about screenplay nominations? Isn't that also a prerequisite for winning Best Picture? In the last six decades, only *The Sound of Music* (1965) and *Titanic* (1997) have won Best Picture without a nomination for either original or adapted screenplay.

All right, which one is it? Which type of nomination – directing, writing, or editing – is most closely correlated with a film's result in the Best Picture category? For that, we need math.

We could just look at the percent of Best Picture winners that were nominated in each of these three categories:

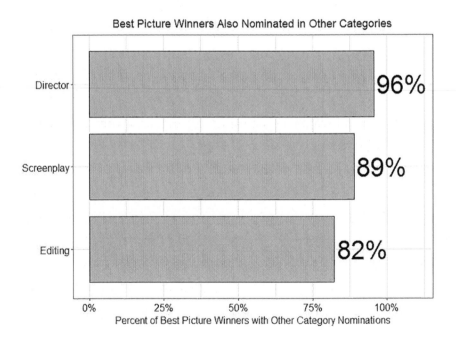

This chart might make it seem like Best Director is most closely related to the Best Picture result. After all, 96% of Best Pictures were also Best Director nominees. But there's a problem with this logic. What if we included a bar for "one hour or longer runtime"? That bar would be at 100% because all Best Picture winners have lasted longer than an hour. *Marty* (1955), a lovely boy-gets-girl-boy-loses-girl-boy-gets-girl-back story, has the shortest runtime of a Best Picture winner at 1:30, certainly longer than 1:00. But does that mean having a runtime longer than an hour is more closely correlated with the Best Picture result than getting a Best Director nomination?

Not quite. The trouble is, 100% of Best Picture *losers* also surpass that one-hour mark. *She Done Him Wrong* (1933) is the shortest nominee at 1:06. Let's say we based all of our Best Picture predictions on whether the film was over an hour long. We'd get every winner right, but we'd get every loser wrong, because we would be predicting that all nominees are winners! Doesn't sound like correlation to me.

Instead, let's look at this a different way: If we made Best Picture predictions based solely on nominations for directing, writing, or editing, how often would we be right? We predict a Best Picture win if the film has the corresponding nomination (directing, writing, or editing) and a loss if the film does not. How often would we guess that a winner won, and how often would we guess that a loser lost?

The first row represents Best Director results for all Best Picture nominees in history. The second row represents Best Film Editing results. The third row covers Best Original and Adapted Screenplay. That's why all three rows are the same length — they are all dealing with the same set of movies, the set of all Best Picture contenders.

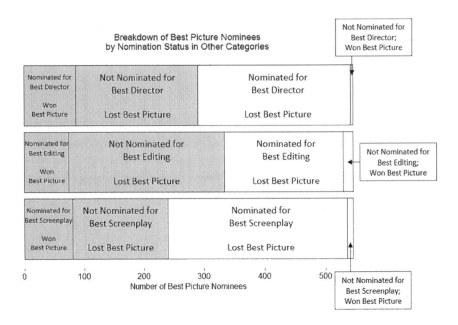

The chart breaks up those Best Picture nominees into four different groups. The first two groups (in grey boxes) represent correct Best Picture predictions: Either the movie was nominated in that row's category (direction, editing, or writing) and won Best Picture, or it wasn't nominated in that row's category and lost Best Picture. The last two groups (in white boxes) represent incorrect Best Picture predictions: Either the movie was nominated for Best Director/Editing/Screenplay but lost Best Picture, or it wasn't nominated for one of those categories but still won Best Picture.

In total, it's the middle row – Best Film Editing – that has the largest amount of grey space. Film Editing dominates in the second box, the one that rewards a category for not nominating movies that go on to lose Best Picture. The reason we want to reward this attribute is that it makes Best Film Editing a very useful indicator of which films to rule out: If a movie doesn't get nominated for Best Film Editing, we're very unlikely to see it win the top prize.

The problem facing screenplay is that it nominates ten movies: five for original and five for adapted. This might help it pick more winners, but it also picks a whole lot of losers.

Are you surprised? As famous as the Director-Picture connection is, and as intuitive as the Screenplay-Picture connection is, it's actually a Best Film Editing nomination that's most closely related to the Best Picture result.

Taking a step back, this makes a lot of sense. More than any other discipline in filmmaking, editing is arguably the tie that binds a movie together. Every other piece of the movie puzzle is only as strong as its editing.

Consider acting: Are the performances in *Network* (1976) by Peter Finch or Faye Dunaway in the famous "I'm mad as hell, and I'm not going to take this anymore" scene half as stunning without the editor's timely cuts between the TV newsman, the studio phone center, and the viewers screaming out their windows? Both Finch and Dunaway won Oscars for their roles.

Consider writing: Does the wildly original backwards-and-forwards plot in *Memento* (2001) make any sense at all without an editor perfectly starting and ending scenes at the precise moments necessary to weave the disparate clips together? Christopher and Jonathan Nolan earned an Oscar nomination for their story.

Examples abound. The sound of *West Side Story* (1961), the cinematography of *Braveheart* (1995), and the visual effects of *The Matrix* (1999) are but three instances of Oscar winners who owe a portion of their victories to the editing department. And with the unique exception of *Birdman*, every Best Director and Best Picture champion ought to send his or her editor a lengthy thank-you note after the show.

To be clear, when I make my annual Oscar predictions, I do not force myself to pick and choose among these different predictors. I use all of them – a directing nomination, a screenplay nomination, an editing nomination, or lack thereof, plus many more inputs – to determine the likelihood that each Best Picture nominee wins. But the better historical predictors, such as Best Film Editing, receive more weight than the ones with weaker track records, such as screenplay nominations.

Although examining non-Best Picture categories is crucial to forecasting the final award of the night, that's not the predictor I'm most often asked about. Much more frequently, people are interested in learning whether fan favorites win more Academy Awards – that is, whether the box office has any bearing on the Oscar race. So let's take a look, shall we?

Chapter 2. Best Visual Effects

Does box office success lead to Oscar wins?

>   MARY:
> George Bailey! Give me my robe!
>
>   GEORGE:
> I've read about things like this, but I've never...
>
>   MARY:
> Shame on you. I'm going to tell your mother on you.
>
>   GEORGE:
> Well, my mother's way up on the corner...
>
>   MARY:
> I'll call the police.
>
>   GEORGE:
> They're way downtown. They'd be on my side, too.

> **MARY:**
>
> Then I'm going to scream!
>
> **GEORGE:**
>
> Maybe I could sell tickets.
>
> —*It's a Wonderful Life* (1946): Nominated for 5 Oscars; won 0

WE BEGAN by exploring just how closely related Best Film Editing is to Best Picture. Next, we arrive at a category that could hardly be less associated with Best Picture: Best Visual Effects. *Wings* (1927), *Ben-Hur* (1959), *Forrest Gump* (1994), *Titanic* (1997), *Gladiator* (2000), and *The Lord of the Rings: The Return of the King* (2003) are the only six movies to win both Best Visual Effects and the night's top prize.

This lack of overlap between Best Picture and Best Visual Effects is hardly surprising. The two categories nominate very different types of movies. Stereotypically, Best Visual Effects is the stomping ground for flashy blockbusters, the highlight of the night for fans of standard Hollywood summer fare. Best Picture, by contrast, tends to nominate smaller-budget critical darlings.

Like all stereotypes, neither of these statements is quite true. Box office smashes *Titanic* and *Lord of the Rings III* won Best Picture, while a few smaller-revenue films like *Ex Machina* (2014) won the visual effects award. But the general idea isn't made up out of thin air. Looking across the years at Best Picture domestic box office takes, as well as Best Visual

Effects since 1991,[1] we can see that Best Visual Effects nominees make more money than their Best Picture counterparts:

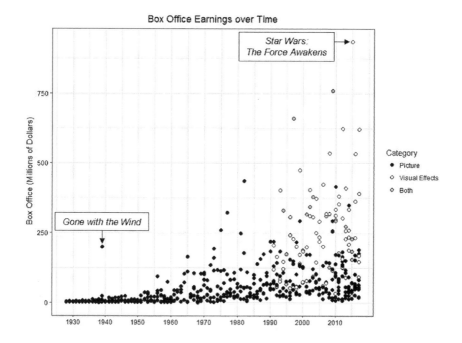

On average, the white circles representing Best Visual Effects nominees are higher on the chart than the black circles representing Best Picture nominees, meaning the Best Visual Effects nominees earned more money.

In both categories, revenues have climbed over time, but that's largely due to the increasing price of movie tickets. Adjusting for inflation, we can redo the above graph on the scale of 2018 dollars:

---

1   Before 1991, Best Visual Effects was not always a competitive award. An award for Engineering Effects was given to *Wings* (1927) for its impressive recreation of aerial warfare, but then the category was retired for a decade. From 1938-1962, the Oscars awarded films for Special Effects, then switched over to the name "Visual Effects" in 1963. From 1972-1990, the award was occasionally honorary, meaning no nominees were announced.

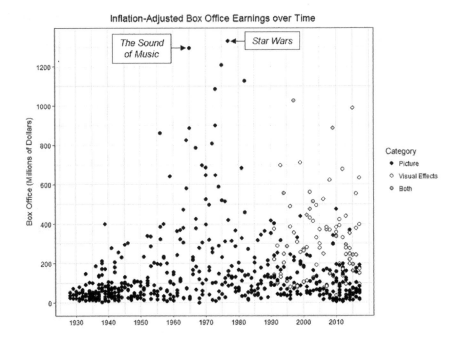

I'm leaving *Gone with the Wind* (1939) off the adjusted-for-inflation graphs, because its total earnings are so high that they don't fit on the chart without making the rest of the data too difficult to see. In other words, *Gone with the Wind* is the highest-grossing Best Picture nominee and winner of all time after adjusting for inflation. Among Best Visual Effects winners, that honor belongs to *Star Wars* (1977), the second-richest film in box office history.

But just because Best Visual Effects *nominates* films that earn more money, does that mean that earning more money helps a film *win* the category? It's a subtle but important distinction: It's theoretically possible that high-earning films are more likely to be nominated in this category, but once the nominees are announced, it might not matter where a film ranks among the nominees.

As an example, in 2014, all five Best Visual Effects nominees (*Captain America: The Winter Soldier*, *Dawn of the Planet of the Apes*, *Guardians of the Galaxy*, *Interstellar*, and *X-Men: Days of Future Past*) surpassed $180 million at the box office, a hefty sum, but it was the least lucrative of those films – *Interstellar* – that went on to win the Oscar.

To answer this question of whether revenue increases Oscar odds, let's once again graph those inflation-adjusted box office numbers over time. But this time, the y-axis (that is, how high or low a point is) will be based on how much more money a nominee made than the average nominee that year. Then we can determine whether films that earn more than their competitors have a better chance of winning. First, the graph for Best Picture:

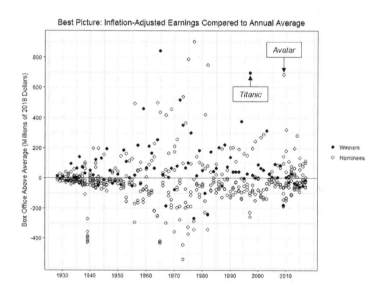

Next, the graph for Best Visual Effects:

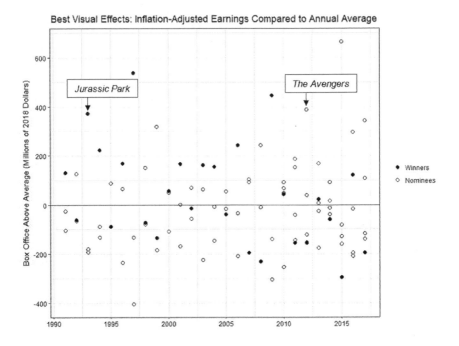

The filled-in circles representing winning films are, on average, just a bit higher than the empty circles representing films that did not win. So, in both Best Picture and Best Visual Effects, once a film is nominated, earning more money boosts its chances of winning.

How much of a boost? Not much. On average, an additional $10 million in revenue translates to a 0.41% increase in Best Picture win probability and a 0.35% increase in Best Visual Effects win probability. It's better than nothing, but these are not huge effects. When predicting the Oscars, I don't include the box office in my model for any category because the data does not show a significant impact of box office receipts on Oscar results.

It is noteworthy that the effects are about the same size for these two disparate categories. Only the biggest blockbusters typically get invited to the Best Visual Effects table, but once there, the added effect of box office earnings on win probability is no larger than the one for Best Picture.

Another reason why this data does not imply that more money causes more Oscar wins: The causation could just as easily go in the other direction. Tens of millions of people tune into the Oscars each year, and every time a film wins an award, it serves as a form of advertising to go see the winning titles, especially for any nominees that are still in theaters. As a result, the numbers above may be inflated for those films that won an Oscar. It's a virtuous cycle for those movies fortunate enough to do well: More attention in theaters may have a small effect on winning Oscars, and winning Oscars can in turn lead to more revenue.

Chapter 3. Best Supporting Actor

What is the right age to win an Oscar?

RHETT:

Did you ever think of marrying just for fun?

SCARLETT:

Marriage, fun? Fiddle-dee-dee. Fun for men, you mean. Hush up. Do you want them to hear you outside?

RHETT:

You've been married to a boy and an old man. Why not try a husband of the right age?

–*Gone with the Wind* (1939): Nominated for 13 Oscars; won 8 (Best Picture, Best Director, Best Actress, Best Supporting Actress, Best Adapted Screenplay, Best Art Direction, Best Cinematography: Color, Best Film Editing)

In 2012, Christopher Plummer held his brand-new Best Supporting Actor Oscar at eye level and exclaimed to it, "You're only two years older than me, darling! Where have you been all my life?" At 82 years old, Plummer had not only won his first Oscar for the ironically named *Beginners*, but also set a record as the oldest winner ever in any of the four acting categories. (The oldest winner in any category is 89-year-old James Ivory for writing *Call Me by Your Name* in 2017.)

Plummer did indeed just miss entering the world before the Oscars did. The latest acting winner born before the first Academy Awards were doled out in 1929 was James Coburn, who took home Best Supporting Actor for playing an abusive father in *Affliction* (1997). Unless a nonagenarian wins an acting Oscar after the publication of this book, Coburn will forever be the last to achieve that distinction.

It is no coincidence that the records for the oldest winner and for the last winner born before the Oscars both belong to the Best Supporting Actor category. What's more, the honor for the oldest nominee goes to Supporting Actor, too: It's once again Christopher Plummer, this time for portraying oilman J. Paul Getty in *All the Money in the World* (2017), replacing Kevin Spacey in the role after sexual assault allegations against Spacey became public.

If you're starting to think that Best Supporting Actor favors older nominees more than the three other acting categories, you're onto something. But in order to prove this, we'll need some math.

For this chapter, I built a model that predicts each acting category for every year in Oscar history using the age of each nominee. For readers with a statistical background, this is a generalized additive model; for all others, that just means that I have created smooth curves for each category in order to make the relationship between winning and age easier to visualize.

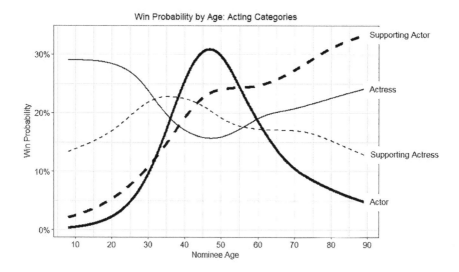

These four curves are clearly quite different, suggesting that the relationship between winning and age varies across the four acting categories. Let's go through them one by one.

Best Actor

This curve is the most dramatic of the four. It rises rapidly until age 47 and then comes crashing back down. The curve's steepness tells us that Best Actor is the most affected by age across the four acting categories. Specifically, it says that actors in their 40s and 50s have a considerably higher chance of winning this award than younger or older actors.

On the left side of the curve, the youngest Best Actor winner is 29-year-old Adrien Brody, who played a Holocaust survivor in *The Pianist* (2002). Brody's win is the lone exception to an otherwise pitiful track record for Best Actor nominees under the age of 30: 1 win, 27 losses. The other three acting categories have each produced at least one winner age 21 or younger, a full eight years shy of the youngest Best Actor winner.

The right side of the graph tells a similarly frustrating story. Best Actor nominees who are 63 and older are just 1-for-22. The lone win went to 76-year-old Henry Fonda in *On Golden Pond* (1981). Fonda has a 14-year lead on the second-oldest champion, John Wayne in *True Grit* (1969).

This pattern – heightened success in middle age – shows just how differently Hollywood treats male and female actors, as we'll see in the next section.

Best Actress

Remember that stat from a few paragraphs ago, that Adrien Brody, at age 29, was the only Best Actor winner under 30? The story for Best Actress could not be more different. Thirty-two winners in this category were in their 20s, from 21-year-old Marlee Matlin (*Children of a Lesser God*, 1986) to 29-year-old Reese Witherspoon (*Walk the Line*, 2005), who was 17 days shy of her 30th birthday.

That's why the curve for Best Actress starts out trending in the opposite direction of the Best Actor curve. The math is saying that it's better to be younger, all the way until around age 47. Coincidentally, the same age that's the high point for actors is the nadir for actresses, providing evidence of the disparity in how men and women are cast.

Of the 90 Best Actress winners, only two were in their 50s: Shirley Booth (*Come Back, Little Sheba*, 1952) and Julianne Moore (*Still Alice*, 2014) were both 54 at the time of their respective wins.

Then the curve picks up again. Seven actresses won the lead category between the ages of 60 and 63, and two were even older: Jessica Tandy (*Driving Miss Daisy*, 1989) was 80, and Katharine Hepburn (*On Golden Pond*, 1981) was 74.

That makes the average age of *On Golden Pond*'s two magnificent winners – Fonda and Hepburn – 75 years old, a full 20 years older than second place in that metric (among films with multiple acting wins). The enemies-turned-partners played by Frances McDormand (60) and Sam Rockwell (49) in *Three Billboards Outside Ebbing, Missouri* (2017) averaged 55 years old.

Supporting Actor

Much like Best Actor, the supporting category for males is unkind to younger entrants. Timothy Hutton broke the mold for his winning portrait of a grief-stricken teenager in *Ordinary People* (1980) at age 20. But aside from Hutton, supporting actors under 27 are 0-for-25. The next youngest winner is 27-year-old George Chakiris as the head of the Sharks in *West Side Story* (1961).

From there, the Best Supporting Actor curve is simple: It just keeps going up. Older is always better in this category, including the examples of Christopher Plummer and James Coburn mentioned at the outset of this chapter. Of the 13 oldest nominees for Best Supporting Actor, 5 of them emerged winners, and all 5 were at least 77 years old.

Sometimes, the Academy uses this slot to honor a career actor who has not yet won, such as Alan Arkin in *Little Miss Sunshine* (2006) – in Chapter 8, we'll examine whether these perennial nominees are more likely to win Oscars. At other times, a character actor takes advantage of one of the numerous Hollywood roles made available to older actors who appear genial or wise, such as Edmund Gwenn as Santa Claus in *Miracle on 34th Street* (1947). So, even if the industry ages actors out of lead roles, there are still opportunities for awards in the supporting category.

Supporting Actress

Yet again, that's not the case for actresses. The Best Supporting Actress curve is the most gradual of the four, meaning that age does not have as large an effect on this category. But it's not exactly horizontal either: The curve goes up until age 35 and then comes back down. So, it's a similar pattern to Best Actor, but not as sharp and with a younger peak.

The curve starts fairly high on the left side in part thanks to three young women who proved that experience is not a prerequisite for delivering a knockout performance. All three of the youngest Oscar-winning actors in history emerged from this category:

- Tatum O'Neal won for playing an orphan and con artist in *Paper Moon* (1973) at age 10

- Anna Paquin won for playing a mute pianist's daughter in *The Piano* (1993) at age 11

- Patty Duke won for playing Helen Keller in *The Miracle Worker* (1962) at age 16

After those three, the trophies pile up in the next few decades of life: 14 women in their 20s, 26 women in their 30s, and 23 women in their 40s have won Best Supporting Actress. The oldest woman to claim this victory was Peggy Ashcroft at age 77 for *A Passage to India* (1984).

The success of a few young supporting actresses may be attributed in part to category placement. Younger actors and actresses are often relegated to the supporting categories, even when they plainly deserve leading nominations. Tatum O'Neal in *Paper Moon* is a great example – she is in basically every scene and steals every one of them as a young thief who can go toe-to-toe in banter with her crooked guardian (played by real-life father Ryan O'Neal). Any objective observer would call the younger O'Neal a lead actress in that film.

But did having a larger role actually help her win an Oscar? Let's find out in Chapter 4.

Chapter 4. Best Supporting Actress

Do bigger roles win more Oscars?

>GEORGE:
>
>Just a second. I'm George M. Cohan. You said you're opening a theater in Philadelphia on July 4th?
>
>ALBEE:
>
>Yes, that's right.
>
>GEORGE:
>
>That's my birthday.
>
>ALBEE:
>
>That isn't why we're opening the theater.
>
>GEORGE:
>
>The salary's all right, but how have you got the nerve to offer us third or fourth billing after my performance tonight?
>
>–*Yankee Doodle Dandy* (1942): Nominated for 8 Oscars; won 3 (Best Actor, Best Musical Score, Best Sound)

FOR THE first eight years of the Academy Awards, there were only two acting categories: Best Actor and Best Actress. The Academy soon realized that this effectively rendered small roles ineligible because most voters picked film stars over cameos.

So, for the ninth ceremony, it introduced new categories for Best Supporting Actor and Best Supporting Actress. Some members weren't thrilled with the idea of putting supporting players on an equal footing with the leads, so for six years the supporting winners were merely given plaques instead of statuettes.

Trophy differences aside, the new supporting categories had their intended effect immediately. Gale Sondergaard, despite receiving eighth billing for her part as a cunning housekeeper in *Anthony Adverse* (1936), won the first award for Best Supporting Actress. It was Sondergaard's first film role after a career spent on the Shakespearean stage, and it set her on a path towards numerous villainous parts over the next decade.

Since that night, however, the Academy hasn't been quite as kind to the truly small parts. Amazingly, all these decades later, Sondergaard still holds the record for the lowest billing of any actress to win an Oscar for a supporting role.[1] On the male side, she has only been beaten once: Mahershala Ali was listed ninth in the final credits for his role as a kindhearted Miami drug dealer in *Moonlight* (2016).

That's not to say that others in the Supporting Actress category haven't had the opportunity to break Sondergaard's record. Another suspicious housekeeper, Helen Mirren's Mrs. Wilson in *Gosford Park* (2001), appears 20th in the credits of Julian Fellowes' murder mystery, the lowest-billed nominee of all time. But Mirren's record comes on

---

1   Hattie McDaniel was also billed eighth in her Oscar-winning role in *Gone with the Wind* (1939), though in that case, the billing was not arranged by prominence in the script. Rather, all the characters based at Tara were listed first.

a bit of a technicality. While most movies list actors in order of importance, *Gosford Park* separated its ensemble cast into the wealthy "Above Stairs" crowd (the first 15 names), the "Visitors" (the next 2), and then the servants "Below Stairs" (the final 35). Dame Mirren came in third in the last group.

On the male side, the lowest-billed nominee is John Huston, listed 15th on *The Cardinal* (1963). That nomination makes Huston, who won Oscars for directing and writing *The Treasure of the Sierra Madre* (1948), one of three Best Director winners to also be nominated for acting in a movie he or she didn't direct.[2] Huston appeared early in *The Cardinal* as an archbishop who mistrusts a young Bostonian priest. After that, his character disappears from the story, but it was enough screen time for Huston to garner the only acting nomination of his career.

Though the category's name includes the word "supporting," 21 actors and actresses (14 male, 7 female) have been nominated despite earning top billing on their films, and three of them won the award:

- Don Ameche: *Cocoon* (1985)
- George Clooney: *Syriana* (2005)
- Patricia Arquette: *Boyhood* (2014)

In *Cocoon*, the primary co-stars were billed alphabetically, so Don Ameche led off by virtue of the letter A. In cases like *Boyhood*, a film shot over 12 years capturing a young man's rise to adulthood, no one person is on screen for a majority of the film, so you could argue that everyone is a supporting actor or actress. If you'd like the more cyni-

---

[2] The others are Robert Redford, the Oscar-winning Best Director of *Ordinary People* (1980), who was nominated for Best Actor in *The Sting* (1973), and Warren Beatty, who won Best Director for *Reds* (1981) and was nominated for Best Actor in both *Bonnie and Clyde* (1967) and *Bugsy* (1991).

cal explanation, some studios are said to deliberately push for their contenders to be mislabeled as supporting to improve their chances, a practice known as "category fraud."

It's no coincidence that these top-billed winners all appeared in more recent Oscar history, while Gale Sondergaard's record for the lowest-billed winner came at the outset. Over the decades, we have witnessed a trend towards more prominent roles for supporting acting nominees:

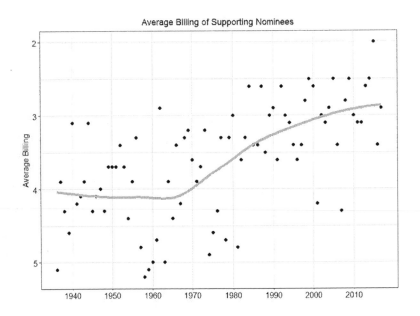

Each point is one year of the Oscars, and the grey curve shows the trend over time. A higher point means that the average nominee was billed earlier in the credits that year.

Indeed, the record for the most prominent total billing (adding up the billing ranks of all five nominees) was set just a few years ago:

2015 Supporting Actor (Total Billing: 8)
Mark Rylance (*Bridge of Spies*): 2 (Winner)
Christian Bale: (*The Big Short*): 1
Mark Ruffalo (*Spotlight*): 1
Sylvester Stallone (*Creed*): 2
Tom Hardy (*The Revenant*): 2

But the record for the least prominent total billing came over half a century ago:

1958 Supporting Actress (Total Billing: 32)
Wendy Hiller (*Separate Tables*): 4 (Winner)
Cara Williams (*The Defiant Ones*): 12
Martha Hyer (*Some Came Running*): 4
Maureen Stapleton (*Lonelyhearts*): 5
Peggy Cass (*Auntie Mame*): 7

Cara Williams had an uphill battle in that race, with such a small part compared to her four competitors. The data for 82 years of Oscar history beginning in 1936 show that every row higher a supporting actress is listed in the credits gains her 1.3 percentage points in her chance to win the Oscar. For supporting actors, the effect is even larger, as higher billing is worth 2.2 percentage points per rank.

Across all years, the average winner's billing is 3.4, while the average losing nominee's billing is 3.7, another way of showing that more prominent roles tend to do a bit better in the supporting categories.

Of course, this is not a hard-and-fast rule. Six actors and six actresses have won the supporting category despite receiving a lower billing than any of their competitors:

Supporting Actor Winners Billed Lowest
Harold Russell (*The Best Years of Our Lives*, 1946): 8
John Mills (*Ryan's Daughter*, 1970): 4
Louis Gossett Jr. (*An Officer and a Gentleman*, 1982): 7
Jack Palance (*City Slickers*, 1991): 6
Jim Broadbent (*Iris*, 2001): 4
Mahershala Ali (*Moonlight*, 2016): 9

Supporting Actress Winners Billed Lowest
Hattie McDaniel (*Gone with the Wind*, 1939): 8
Mercedes McCambridge (*All the King's Men*, 1949): 5
Gloria Grahame (*The Bad and the Beautiful*, 1952): 6
Jo Van Fleet (*East of Eden*, 1955): 6
Maureen Stapleton (*Reds*, 1981): 7
Judi Dench (*Shakespeare in Love*, 1998): 6

Both the first man and the first woman on these lists overcame the odds in more ways than one. Not only did they win Oscars with small roles, something that math says is less likely than winning the award with a more prominent role, but they also had to deal with daunting off-screen challenges.

Harold Russell lost both arms while serving in the U.S. Army during World War II and then portrayed on-screen a returning veteran who suffered the same injury. The performance was so moving that the Academy made it the only role to ever be recognized twice – once as an honorary Oscar and once as the winner for Best Supporting Actor, with both awards presented on the same night.

Hattie McDaniel made Academy history by becoming the first black actor or actress to win an Oscar for her role as Scarlett's slave in *Gone with the Wind*. Combating racism in Hollywood throughout her life, she managed not only to land the part but also to deliver one of the greatest performances of all time, lead or supporting. It's no easy feat to steal multiple scenes of a movie that features many of its era's most legendary talents, yet McDaniel did just that in star-studded *Gone with the Wind*.

The Academy has developed a bias over the last half-century towards nominating supporting roles that claim more screen time, a bias that did not exist in the early decades of the Oscars. And once the nominees reach the ceremony, more prominent roles enjoy a slight edge. But as Russell, McDaniel, and ten others have proven, there may be small parts, but deliver a strong enough performance and you could still wind up on top.

Chapter 5. Best Makeup and Hairstyling

Does the total number of nominations matter?

> CLOWN:
> Why the rehearsal just for makeup, Brad?
>
> BRADEN:
> Because you left off your eyelashes last performance last season.
>
> –*The Greatest Show on Earth* (1952): Nominated for 5 Oscars; won 2 (Best Picture, Best Story)

THE MOMENT the Oscar nominations are released, pundits rush to anoint favorites. Within seconds, whichever film received the most nominations is suddenly considered the frontrunner for Best Picture, and often for every other category in which it's nominated.

Is this correct? If so, how big of an advantage is it to have more nominations than any other film? Let's look at this through the lens of two categories on opposite extremes: Best Makeup and

Hairstyling – a category that honors a very specific aspect of a film – and Best Picture – a category that pays tribute to an entire film.

The reason we need to do the analysis twice is that the stories behind the data could be very different. Best Makeup and Hairstyling should, in theory, be unrelated to nearly every category other than Best Picture. While cosmetics are one of the many factors that go into making a great movie overall, there's no particular reason to believe that a Best Supporting Actor nomination, for example, makes a win for makeup and hair more or less likely.

You could argue that there might be a tenuous connection between makeup/hair and Best Director, Best Costume Design, or Best Production Design, but in truth, those three categories really weren't created to evaluate makeup or hair.

Best Picture, on the other hand, should be connected to every single other category. A nomination in any category up or down the ballot should signify that a film is strong in that particular area, leading to a better movie on the whole. But of course, what "should" happen isn't necessarily what does happen. So let's find out how Oscar voters actually behave.

In 1980, *The Elephant Man* was nominated for eight Oscars, including Best Picture, but won none of them despite its incredibly moving portrayal of a man with significant deformities and the doctor who was the first to show him human kindness. It must have been a frustrating night for the film's backers: Only 13 movies have ever scored eight or more nominations without winning a single one, from *The Little Foxes* (1941) to *American Hustle* (2013).

Amidst its shutout, *The Elephant Man* surely deserved one Oscar: Best Makeup and Hairstyling, for making the titular character's condition look so lifelike. The trouble is, that award didn't exist yet. Realizing that such extraordinary makeup work sorely deserved a trophy, the Academy established the category the following season.

In 1981, both of the inaugural nominees – *An American Werewolf in London* and *Heartbeeps* – weren't nominated for anything else. It's fortunate for those films that the category was created just in time, or they wouldn't have been invited to the Oscars at all. But it doesn't help us on our mission to learn whether there's a relationship between nomination count and the Best Makeup and Hairstyling Oscar because these two films tied on nomination count.

The following year, 1982, provides our first evidence. The two nominees were *Quest for Fire*, nominated for nothing else, and *Gandhi*, with 11 total nominations. *Gandhi* is a masterpiece of a biography, detailing in epic fashion the life and death of the pacifist revolutionary. The dramatic funeral scene still holds the record for most extras used in a single movie scene. *Gandhi* went on to win eight Oscars including Best Picture, one of only 11 films to earn more than seven Oscars on more than ten nominations. But absent from the win tally was Best Makeup, which went to *Quest for Fire*.

Our first piece of evidence suggests a negative correlation between nomination total and Makeup and Hair. But one year doth not a trend make. Let's examine all years at once:

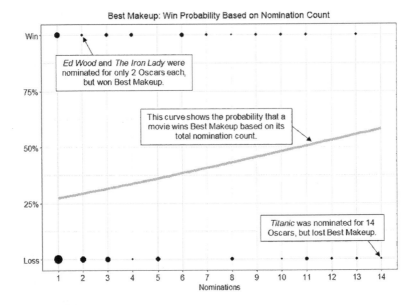

The points along the top of the graph represent films that won Best Makeup and Hairstyling. The points at the bottom lost. The bigger the point, the more movies that point represents. The farther to the right a point is, the more total nominations it received. For example, that tiny point in the bottom right represents *Titanic*, which accrued 14 total nominations (which is why the point is on the right side), is the only Best Makeup and Hairstyling nominee to ever earn that many total nominations (which is why the point is so small), and lost the makeup category (which is why the point is on the bottom row instead of the top).

The diagonal line across the middle represents the probability that a film with a given number of nominations won this category, using a classic intro-to-statistics method called logistic regression, which basically answers the question of how likely an event is to occur (in this case, winning Best Makeup) based on some other data (the total number of nominations). The line does slope upwards, though it's not all that steep. This means there is a positive relationship between total nomination count and winning Best Makeup, but it's not that strong.

This analysis has one problem, though. This chart ignores the fact that movies are grouped by year, and each film only competes against others in that year.

To illustrate the point, consider the 2000 Oscars. I'll start out by telling you that the Makeup and Hairstyling winner was *How the Grinch Stole Christmas*, which had a total of three nominations. Armed with only that information, this sounds like a data point that favors the idea that large nomination totals don't help a nominee's chances in this category. After all, three isn't a very big number.

But the other two makeup/hair nominees that year, *The Cell* and *Shadow of the Vampire*, received just one and two nominations, respectively. So actually, 2000 does support the hypothesis that more nominations help because the movie with the most nominations won.

On the opposite end of the spectrum is 1998: *Elizabeth* won the makeup Oscar and had 7 total nominations. Sounds like a fair amount, right? But it beat out *Shakespeare in Love* (13 nominations) and *Saving Private Ryan* (11 nominations), so 1998 was actually a data point that contradicts the idea that more nominations are a good thing.

To fix the analysis, we'll reevaluate every movie, not on the basis of the total number of nominations it had, but according to how many it had relative to its competition. First, we'll calculate the average for each year. Then, we'll calculate how many more or fewer nominations each contender had compared to that average.

Let's use 1995 as an example. The three Makeup and Hair nominees were *Braveheart* (10 total nominations), *My Family* (1 nomination), and *Roommates* (1 nomination). The average of these three numbers (10, 1, and 1) is 4. So, *Braveheart* gets a score of 10 − 4 =

+6. The other two get a score of 1 − 4 = −3. Note that the three scores (+6, −3, and −3) add up to 0, which will always be the case.

In other words, in 1995, *Braveheart* scored 6 more nominations than the average nomination total of the three Best Makeup nominees. *My Family* and *Roommates* each had 3 fewer total nominations than the average for this category. *Braveheart* went on to win the makeup award.

We can then perform this analysis for every year individually from 1981 through 2017, and what we find is that, in general, a movie is more likely to win Makeup and Hairstyling when its total nominations exceed the average total nominations of the films nominated in the makeup category. That likelihood increases as the number of nominations above average increases.

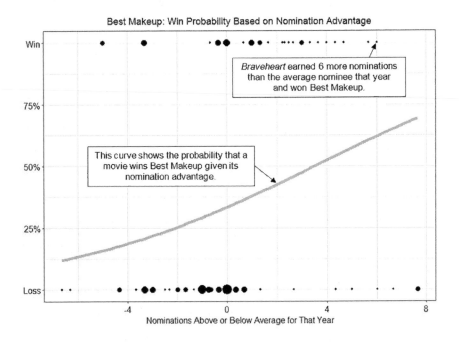

Once again, the larger the dot, the more movies represented. The diagonal line represents the probability that a movie with that many nominations above or below average will win Best Makeup.

This time, the curve is steeper, meaning we have even stronger evidence that earning more nominations helps a movie win this award. The nominees on the far left, the ones with very few nominations relative to their competitors, have less than a 20% chance to win the Oscar. The nominees on the far right exceed 60%. That's a big difference.

As a matter of fact, this is evident just from looking at the points. It's pretty clear that most of the points on the top row are bunched together on the right side. The two biggest outliers, each with scores of -5, are *Quest for Fire*, the one that beat *Gandhi*, and *Mrs. Doubtfire* (1993), which toppled multi-nominated *Philadelphia* and *Schindler's List*. It's no wonder that *Mrs. Doubtfire* broke the trend: Without impressive makeup work, Robin Williams wouldn't have been nearly so believable as a female housekeeper.

What's going on here? Why do films with more nominations have such a strong leg up in the Best Makeup and Hairstyling race? Is it really that the films with better cinematography or acting or editing just happen to have the best makeup?

That's certainly possible. If a film has a talented director or producer, it stands to reason that just about every aspect of the film is likely to shine. On top of that, I'd posit a couple of other theories. If a movie receives a lot of nominations, there may be voters who want to honor it elsewhere on the ballot, but due to stiff competition in those categories instead support the movie for Makeup and Hair. Or, if there's one movie a voter strongly prefers, he or she may choose to vote for it in as many categories as possible.

And in some cases, voters don't actually get around to seeing all of the nominees: Across all years, an average of 54.9 films receive at least one nomination per year. Academy members are busy people like the rest of us, with families and jobs and to-do lists. Seeing over 50 films throughout January and February might not be realistic for everyone. Some choose to abstain from categories in which they haven't judged all of the options. But others may vote anyway – it's impossible to police whether a voter has seen every nominated film – so their votes are much more likely to go to movies they've actually seen, which tend to be the more nominated ones.

Best Picture, on the other hand, doesn't encounter that problem. Nearly every voter has seen all of the nominees. So let's see what happens when we apply the same method to the evening's final category:

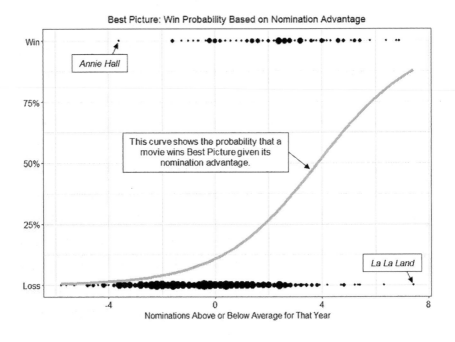

It turns out the pundits have nailed it on this one. There's a strong correlation here. Using all 90 years of Oscar data, only one of the 144 Best Picture nominees with a score below -1.6 has ever won Best Picture: *Annie Hall* (1977). Woody Allen's fourth-wall-breaking romantic comedy won despite garnering only 5 nominations, defeating three competitors that each gathered 11 nods: *Julia*, *Star Wars*, and *The Turning Point*.

On the other end of the spectrum, 27 films have a score above +4.2, and 16 won Best Picture. Yes, that leaves room for 11 losses – most spectacularly, *La La Land* (2016) and its 14 nominations lost despite a score of +7.4, and *The Song of Bernadette* (1943) couldn't win even with a +6.3. But still, 16 / 27 is a 59% winning rate, which is pretty darn good in a race that only 16% of nominees have won. When I predict the Oscars, I use this information on how many nominations in other categories a film has, though I don't treat all nominations equally. (When predicting Best Picture, for instance, a Best Director nomination is more predictive than a Best Visual Effects nod.)

With that, back to the original statement: The film with the most nominations is the favorite. Let's test that. For Best Makeup and Hairstyling, 25 of the 36 winners had the most total nominations among the contenders (or were tied for the most). For Best Picture, 58 of the 90 champions have led the nominations announcement.

It turns out that whether we're looking at major categories like Best Picture or artistic races like Best Makeup and Hairstyling, the verdict is the same: the more nominations, the higher the chance of winning the Oscar. On this Oscarmetrics question, common sense and conventional wisdom reign supreme.

Chapter 6. Best Costume Design

Which guild is the best Oscar predictor?

> LOLLIPOP GUILD:
> We represent the Lollipop Guild, the Lollipop Guild, the Lollipop Guild. And in the name of the Lollipop Guild, we wish to welcome you to Munchkinland.
> –*The Wizard of Oz* (1939): Nominated for 6 Oscars; won 2 (Best Original Score, Best Original Song)

EVERY FEBRUARY, with the Oscars just around the corner, I spend one Tuesday night refreshing Twitter. I have most of the necessary data, but my spreadsheet for Best Costume Design still has one big blank column for the Costume Designers Guild Awards.

These awards aren't popular enough to qualify for television coverage, forcing me to use Twitter for results. But they're very important in the world of Oscar prediction because they provide insight on a category – Best Costume Design – that has relatively few other predictors. Only the BAFTAs, the Critics' Choice Awards, and a few other lesser-known awards have a costume design category, some more predictive than others.

In fact, in 2002 and 2003, the Costume Designers Guild (CDG, the union representing costume designers) was the only one of these three groups to pick the correct winners. The Critics' Choice Awards didn't add a costume category until 2009, and the BAFTAs chose *The Lord of the Rings: The Two Towers* (2002) and *Master and Commander: The Far Side of the World* (2003). The CDG got it right both times, with *Chicago* (2002) and *The Lord of the Rings: The Return of the King* (2003).

But just because the Costume Designers Guild has outsized importance in predicting Best Costume Design at the Oscars, that doesn't mean it's the best predictor of its corresponding category compared to other Hollywood guilds that host award ceremonies.

For one thing, the guild has only been handing out awards since 1998, so it has less of a track record than other guilds to prove its worth as a predictor. Here is a timeline showing when the Oscar-precursor guilds began handing out awards:[1]

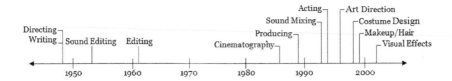

For another thing, the Costume Designers Guild gives itself an unfair advantage. At first glance, when I tell you that the CDG has picked 9 Oscar winners in its 20 years of existence, that doesn't sound too bad. But unlike most guilds, the CDG splits its awards by genre. It started off with just one category in 1998, when 1950s-era social commentary *Pleasantville* won the inaugural prize (an incorrect Oscar prediction, as *Shakespeare in Love* took the costume Oscar that year).

---

1   The Make-Up Artists and Hair Stylists Guild started giving out awards in 1999, but then stopped after 2003. The makeup and hair awards resumed a decade later, in 2013.

But the next year, the costume designers split their awards, one for contemporary and one for period/fantasy. In 2005, period and fantasy became two separate categories, bringing us up to three.

Now, that 9 out of 20 sounds less impressive. Think about it this way: If you're trying to roll a 6, and I only give you one die, you have a one out of six shot, about 17%. But if I give you three dice, your chance of getting at least one 6 jumps up to 42%. Taking three shots is always better than taking one.

Therefore, to determine which guild has historically done the best job of predicting the Oscars, we're going to need a scoring system that fairly accounts for all of this. This is a crucial question in Oscar prediction because we want to give more weight to guilds that do a better job of forecasting the Academy Awards.

In total, we have 12 entrants in our Battle of the Guilds. There are no guild awards for Best Score, Best Song, Best Foreign Language Film, or the three short film categories (Live-Action Short, Documentary Short, and Animated Short).

A few of the guilds count for multiple categories: The Producers Guild confers awards for feature films (i.e., Best Picture), documentary features, and animated features. The Screen Actors Guild serves as a predictor for all four acting categories. The Writers Guild predicts both Best Original Screenplay and Best Adapted Screenplay.

In order to handle all of these different types of awards, we'd like to invent a scoring system that follows a number of rules:

- No matter how many selections a guild makes, correctly predicting the Oscar is always worth positive points, and an incorrect prediction is always worth negative points.

- Picking the winner out of n selections always grants a guild more points than picking the winner out of $n+1$

selections. In other words, we're rewarding guilds that get it right in fewer guesses.

- For higher numbers of selections, the differences get smaller. Picking the winner with just one guess is significantly harder than picking the winner with two guesses. But picking the winner with nine guesses is only barely harder than picking the winner with ten guesses.

- A guild that only nominates one film every year (meaning the guild just hands out an award without announcing nominees in advance), and gets exactly half of the Oscar winners right over time, will end up with a final score of 0 points. This is considered to be an exactly average performance.

- The maximum attainable score for any guild in any year is 100 points, if it only nominates one film and that film wins both the guild and the Oscar. The choice of 100 is arbitrary – the number we choose won't affect the final standings.

- For guilds that have above-average performances, meaning they get more right than wrong, their scores will be higher if they've participated in more years. This penalizes newer guilds to avoid the case in which a new guild opens up, gets one prediction right in its first year, and is suddenly named the winner of this battle.

I have created a formula for this chapter that fits all of these requirements. For those who'd like to skip the math and go straight to the results, just keep on reading. For the more curious among you,

I include an addendum at the end of this chapter which lays out the formula, along with a specific example (the 2016 Costume Designers Guild Awards) to illuminate what's going on under the hood.

Most guilds are rather straightforward to score. For example, the Cinema Audio Society awards one film for best sound mixing of the year, and all we have to do is compare that result to the Oscar for Best Sound Mixing. In three cases, there are ambiguities, so I chose the interpretations that are most generous to the guilds:

- The Visual Effects Society (VES) has two primary awards for live-action films: one for visual effects in an effects-driven movie and the other for supporting visual effects. The first category is a strong predictor of the Academy Award for Best Visual Effects because the Oscars are far more likely to give this award to a movie that heavily relies on visual effects. In fact, only one Oscar winner (2011's *Hugo*) has ever emerged from the VES's supporting effects category, so I removed the supporting category from the data. This makes the VES look worse in 2011 by not giving it credit for the *Hugo* win, but it makes the society look better in every other year since it considers the guild to be picking correct Oscar winners on one guess instead of two.

- The Make-Up Artist & Hair Stylists Guild Awards, for being around for a relatively short time, have a remarkably high number of ever-fluctuating categories. There are separate awards for hair and makeup, and the category names change frequently. The Oscars have a far greater overlap with the makeup categories than with the hair categories; in fact, no Oscar winner has ever won or been nominated for any hair category without also being recognized by the makeup categories. So, I only included the guild's makeup categories for period films, character makeup, and special effects.

- The Motion Picture Sound Editors (MPSE) have changed their categories over the years. My solution: For the 1960s, when the MPSE's Golden Reel Awards were young, I use their Feature Film award. Afterwards, I use their Sound Effects award. This penalizes the MPSE in a handful of years when the Oscar winner won a parallel MPSE category (such as 2004, when *The Incredibles* won MPSE's animated category), but it helps the MPSE in most years. I only count years when the Oscars nominated multiple films for Best Sound Editing, and I give the MPSE full credit for picking *Skyfall* in 2012, even though that film tied at the Oscars with *Zero Dark Thirty*.

Performing my calculation for every guild for every year, this chart shows us the final standings:

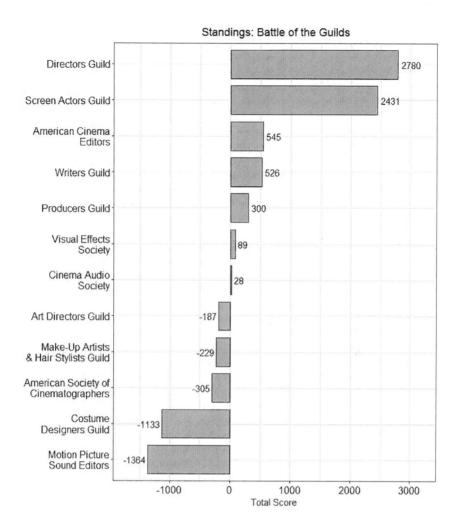

Ladies and gentlemen, we have a winner. Put your hands together for the Directors Guild. Amazingly, the Directors Guild has never once failed to nominate the eventual Best Director winner at the Oscars, save for the inaugural year, 1948, when the two organizations were on different calendars and *The Treasure of the Sierra Madre* missed out on a Directors Guild nod. Even more impressively, the Oscar winner has also won the Directors Guild in all but seven years, as I'll discuss in more detail in the Best Director chapter.

The Directors Guild is a worthy winner indeed, but close on its heels is the Screen Actors Guild (SAG). Marcia Gay Harden (Best Supporting Actress, *Pollock*, 2000) and Christoph Waltz (Best Supporting Actor, *Django Unchained*, 2012) are the only two acting winners at the Oscars to ever miss out on SAG nominations. And of the 96 SAG acting champions in history (24 years times 4 acting categories), a hefty 74% of them won an Oscar.

The next trio features some of the most prolific awards on the guild scene. The Eddies (presented by the American Cinema Editors guild) have been around since 1961, while the Writers Guild and Producers Guild serve as precursors to two and three Oscar categories, respectively. As long as an organization gets more than half of its picks correct, my formula is designed to favor those that have handed out more awards, and that's what we see here.

In the middle tier, the Visual Effects Society, the Cinema Audio Society, the Art Directors Guild, the Make-Up Artists & Hair Stylists Guild, and the American Society of Cinematographers all perform okay. These scores mean they're good enough to be considered when making Oscar predictions, but unlike the awards higher up on the list, you'd feel a whole lot better if some other indicators confirmed their picks before checking off their winners in your Oscar pool.

Finally, at the bottom, we have the Costume Designers Guild and the Motion Picture Sound Editors. The costume designers got off to a rough start, not even nominating the Oscar winner in each of their first four years. They finally picked a winner with *Chicago* (2002). Since 2002, their success has picked up a little, and perhaps somewhere down the road the CDG will climb above zero in this metric.

The Sound Editors, to be fair, were somewhat between a rock and a hard place in this competition. As I mentioned above, either I included all MPSE categories, making the MPSE look worse for taking more dart throws per year than other guilds, or I included fewer MPSE categories, meaning some of the MPSE's correct picks don't count. I chose the interpretation that most benefitted the guild, but it wasn't enough to move the MPSE beyond last place in this ranking. And this is an accurate reflection of what it's like trying to use their Golden Reel Awards to predict the Oscars. With categories ranging from Sound Effects to ADR (Automated Dialogue Recreation) to Animated to Foreign, it's not always clear what Oscar signal the MPSE is sending.

None of this makes a guild better or worse at rewarding talent. If anything, it's perhaps more likely that the Academy misses the mark; presumably, a guild's members know their craft better than anyone else. When a guild disagrees with the Academy, it doesn't mean the guild made a mistake. It just means we have to give that guild less weight in order to win those Oscar pools.

• • •

For those who are interested, here is my scoring system that follows all of the rules set out earlier in the chapter: If the guild makes n selections and one of them matches the Academy, the guild earns $100/2^n$ points. If the guild makes n selections and all of them are wrong, the guild loses $100 * (1-1/2^n)$ points. The guild can gain or lose points both for nominating the Oscar winner and for awarding the Oscar winner.

This looks complicated on paper, but it will make more sense when we apply it to a real-life example: the 2016 Costume Designers Guild Awards.

First, we look at whether or not the Oscar winner that year (*Harry Potter* spinoff *Fantastic Beasts and Where to Find Them*) was nominated by the guild. In this case, it was. In fact, a total of 15 films were nominated by the CDG, 5 per category across the contemporary, fantasy, and period categories. *Fantastic Beasts* scored a nomination in the fantasy category. So, the CDG gains $100/2^{15} = 0.003$ points for this correct nomination.

Second, we want to know whether or not *Fantastic Beasts* won an award from the CDG. In this case, it did not, as *Fantastic Beasts* lost the fantasy category to *Doctor Strange*. The Marvel product was one of three incorrect predictions the CDG made that year, along with *La La Land* for contemporary film and *Hidden Figures* for period film, so the CDG loses $100 * (1-1/2^3) = 87.5$ points.

Overall, the CDG's score for that year is $0.003 - 87.5 = -87.497$. That's not a great showing. The CDG gets very little credit for nominating the correct winner because it gave itself 15 chances to get it right. And the CDG gets a lot of negative credit for failing to award the winner because it gave itself three tries and still got it wrong.

Chapter 7. Best Production Design

Which genres win more Oscars?

> ADRIANA:
> I'm from the twenties and I'm telling you the golden age is La Belle Époque.
>
> GIL:
> I mean, and look at these guys. I mean, to them their golden age was the Renaissance. You know, they'd rather — you know they'd trade Belle Époque to be painting alongside Titian and Michelangelo. And those guys probably imagine life was a lot better when Kubla Khan was around.
>
> –*Midnight in Paris* (2011): Nominated for 4 Oscars; won 1 (Best Original Screenplay)

I'VE TRAVELED a long time ago, in a galaxy far, far away.

I've clashed swords with Robin Hood, raced chariots with Judah Ben-Hur, fought a Roman garrison with Spartacus, planned attacks with Lawrence of Arabia, and marched for freedom with Gandhi.

My heart pounded when I was trapped with a fading Hollywood star, when I fooled the mob, when I inherited an evil enterprise from my father, when I found the Holy Grail, when I destroyed the one ring to rule them all, when I survived the Spanish Civil War, when I tried to warn the Na'vi, when I hung from a Parisian clock tower, when I was framed for murder in a grand hotel, and when I silently saved the fish I loved.

I've sung with an American in Paris, a Welsh teacher in Bangkok, a matchmaker in New York, a cockney girl in high society, a just king at a round table, and a murderess in Chicago.

I've danced with gangs on the West Side, street urchins in London, cabaret singers in Berlin, showgirls in fin-de-siècle Paris, and dreaming commuters in Los Angeles.

I've witnessed awful terrors: the Civil War in Georgia, a jewel-crazed husband slowly driving me insane, cruel racial injustice in Alabama, the Holocaust, and an unsinkable boat dooming hundreds to an icy death.

But from the depths of despair, I've seen the potential of humanity in a union on the waterfront, in an inspirational yet controversial general, in a pair of crusading journalists, in a genius composer, in a brilliant bard in love, and in our nation's greatest leader.

I am indeed a well-traveled man, through both space and time, all from my comfortable seat at my local theater. These 38 films I referenced genuinely made me feel like I was in another place or era, at least for a couple of hours – and for the many days I spent thinking about the movies after I had seen them. It takes a full cast and crew to transport us, but arguably more than anyone else, it takes a top-notch production design team.

This team has the job of taking a set, which often begins as a barren soundstage in the Los Angeles area, and transforming it to the point that it feels like the movie was truly filmed in the period and place that the action occurs in. The 38 films that opened this chapter did such an incredible job of recreating moments in time and space that all 38 deservedly won Oscars for Best Production Design, or its predecessor, Best Art Direction.

In theory, all artists in this category can do is enliven the scripts that they are given. They don't write the screenplays; they don't pick the genre, the setting, or the era. But do Oscar voters have a habit of awarding those who work on certain types of movies?

To answer this, I went through every Best Production Design nominee since 1967, the year that Best Art Direction became one category of five nominees, as it has been ever since (prior to that, there were 25 years in which black-and-white and color films had separate art direction categories). For each movie, I determined the genre, location, and year the plot took place. I recognize that many readers (myself included) have a particular interest in the Best Picture race, so for this chapter and many others, I also include data from that category.

Admittedly, a number of these classifications had to be subjective. Was 2012 Best Production Design winner *Lincoln* a war film or a biographical film? Though the opening scene memorably takes place on a brutal battlefield, the majority of the subsequent scenes take place in Washington, D.C., far from the fighting, so I decided it was primarily biographical. For many films, I applied the rule that more specific genres took precedence over less specific ones, so *The Natural* (1984) is a sports film, though it could also qualify as a fantasy or a drama.

Classifying location was hardly cleaner, since I decided to choose just one setting for each film to make the subsequent analysis more interpretable. *Forrest Gump* (1994) tells his story from a bus stop in Savannah, Georgia; travels to Washington, D.C.; fights in Vietnam;

and runs through at least a dozen states as he crisscrosses America. But I decided the film was primarily set in Alabama, the state where Forrest is raised by his mother, where Forrest stars as a running back in college, and where Forrest and Jenny reunite.

Then there are ambiguities of time. In *Camelot* (1967), the first winner of Best Production Design under the modern category structure, Richard Harris bellows, "This is the time of King Arthur, when we shall reach for the stars! This is the time of King Arthur, when violence is not strength, and compassion is not weakness!" But what time is that? Though Arthur may very well be only a legend, a tenth-century Welsh document claims that the famous British king met his end in 537, so that was the year I chose.

Other films take place over many years, such as 2008 Production Design champion *The Curious Case of Benjamin Button*, a curious case for choosing a year when the plot involves a man growing younger over the course of the twentieth century. I chose to mark it down as 1918, the year that Benjamin Button is born and the year that a backwards-running clock presumably curses the newborn baby to his reverse life.

While every movie is assigned a genre, I did not mark down a time and/or place for a small number of them. A couple of examples:

- *The Grand Budapest Hotel*, Wes Anderson's delightfully quirky comedy that won this category in 2014, explicitly takes place in the Republic of Zubrowka, a fictional country that doesn't fit within the bounds of actual geography.

- There are various estimates for when the *Lord of the Rings* trilogy takes place – including some vague ones by J.R.R. Tolkien himself – but in general I don't believe those movies are meant to be placed on an actual timeline of real-world events.

Caveats aside, here are the Oscar results by genre, setting, and time period, going back to 1967 for Best Production Design and the inaugural ceremony for Best Picture:

Genre

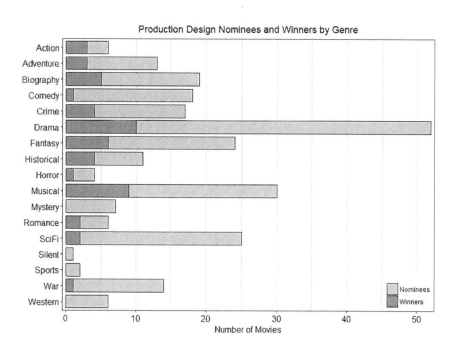

I'm playing fast and loose with the term "genre" here because silent films are less a genre than a style. The same goes for animated films, which I classify separately in the Best Picture section. That's irrelevant here, as no animated film has been nominated for Best Production Design, unless you include partially animated films like *Mary Poppins* (1964), *Bedknobs and Broomsticks* (1971), and *Who Framed Roger Rabbit* (1988).

By far the most nominees come from the drama genre, which along with comedy serves as a default category for movies that don't quite fit into any of the other groupings. A fair number of musicals, science fiction movies, and fantasies also make the cut.

On a percentage basis, action films have proven the most successful, with three of the six nominees coming out on top: *Batman* (1989), *Crouching Tiger, Hidden Dragon* (2000), and *Mad Max: Fury Road* (2015).

Historical films are up next, with a 36% success rate thanks to wins by 4 out of 11 nominees: *Nicholas and Alexandra* (1971), *All the President's Men* (1976), *Schindler's List* (1993), and *The Madness of King George* (1994).

On the other end of the spectrum, science fiction, war, and comedy films have struggled to win this award. Despite 57 nominations across these categories, only 4 movies have taken the Production Design Oscar. Two of them were sci-fi, *Star Wars* (1977) and *Avatar* (2009); one was a war film, *The English Patient* (1996); and one was a comedy, *The Grand Budapest Hotel* (2014).

Four genres in this dataset have never won: mystery, silent, Western, and sports. That said, prior to 1967, silent films *The Dove* (1927), *Tempest* (1928), and *The Bridge of San Luis Rey* (1929); Western *Cimarron* (1931); and boxing film *Somebody Up There Likes Me* (1956) all won Oscars for Art Direction.

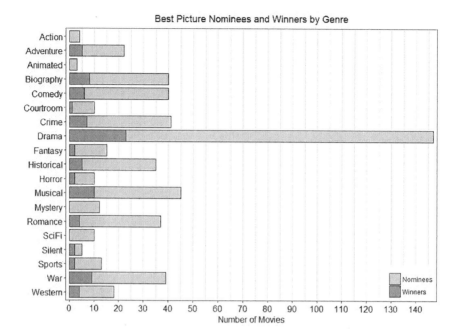

The first result that jumps out from the Best Picture graph is that far more dramas make the cut. All other genres are relegated to secondary status.

Silent films technically have the best winning percentage, with *Wings* (1927) and *The Artist* (2011) making it two-for-five, or 40%. But the sample is so small that it doesn't mean much. No other genre reaches the 25% threshold, though war (23%), adventure (23%), musical (22%), and Western (22%) all get close.

Action (0/4), animated (0/3), mystery (0/12), and science fiction (0/10) haven't fared nearly so well. Of course, classifying genre isn't an exact science. You could argue that *The Lord of the Rings: The Return of the King* (2003) and *The Shape of Water* (2017) are sci-fi, but I would counter that they are more closely aligned with fantasy. Some might consider *Rebecca* (1940) or *In the Heat of the Night* (1967) to be mystery winners, though I decided that Hitchcock's psychological thriller *Rebecca* was closest to horror among these genres, and *In the Heat of the Night*'s detective story made crime a better fit.

Even action may have had a few winners since the boundary between action and adventure is so subjective. Animated films, having lost for *Beauty and the Beast* (1991), *Up* (2009), and *Toy Story 3* (2010), are the one category that has undeniably never won Best Picture.

Setting

The entertainment industry is dominated by New York and California: 16% of all Best Picture nominees take place in the Empire State, and 8% take place in the Golden State. Here is the map of how many nominees for Best Production Design occur in each state:

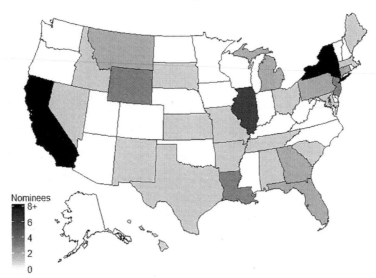

Next, here is the same map for Best Picture:

Best Picture Nominees by State

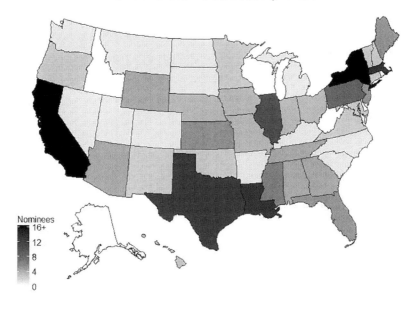

Though New York and California regularly appear on the Best Picture nomination lists, the winners' circle is a very different story. New York is doing just fine, with 16 winners to its name out of 90 nominations, good for a respectable 18% clip. California, however, is just 2-for-41, with only back-to-back Los Angeles winners *Million Dollar Baby* (2004) and *Crash* (2005) carrying the mantle for the home of Hollywood. That makes California arguably even unluckier than the District of Columbia, which has a dozen Best Picture nominees – including my favorite movie *Mr. Smith Goes to Washington* (1939), *The Exorcist* (1973), *All the President's Men* (1976), *A Few Good Men* (1992), and *Lincoln* (2012) – but is still searching for that first win.

After New York, the next most popular Best Picture-winning locations are Illinois with four – *The Sting* (1973), *Ordinary People* (1980), *American Beauty* (1999), and *Chicago* (2002) – and Florida with three – *It Happened One Night* (1934), *The Greatest Show on Earth* (1952), and *Moonlight* (2016).

Six states have never hosted a Best Picture nominee: Alaska, Delaware, Idaho, Rhode Island, West Virginia, and Wisconsin. This is not to say that no Best Picture nominee has ever stopped by these states. For example, Alvy Singer travels to Wisconsin to meet his girlfriend's family for an awkward dinner in *Annie Hall* (1977). But it's clear that the action is primarily set in New York, and I only designated one state per movie.

Zooming out, here is the map of Best Picture nominees by country:

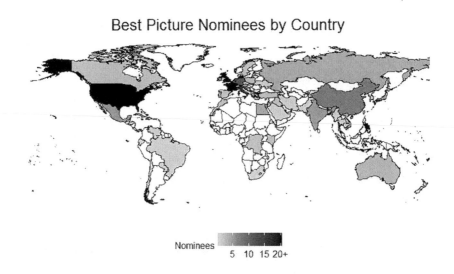

Best Picture Nominees by Country

A hefty 73% of all Best Picture nominees take place in America, Britain, and France – 54% in America alone. Among countries with at least ten nominees (America, Britain, France, Italy, and Germany), the winning percentages are all between 15% and 20%, so there does not appear to be a bias towards any particular country. For nominations, on the other hand, there is a Western tilt, given that all of the countries just listed are in America or Western Europe.

Time Period

This graph shows the year the action takes place in Best Production Design nominees:

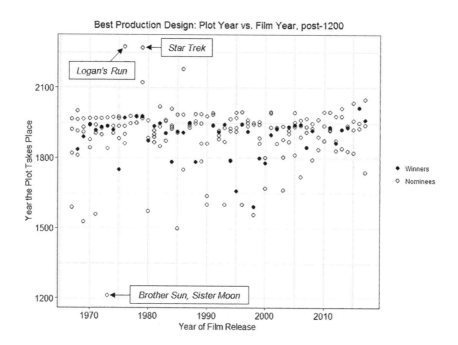

To make the graph more readable, I left off two points that occur far earlier than the year 1200. Since 1967, the Best Production Design nominee set at the earliest time is *Gladiator* (2000), which falls in the second century. The next earliest nominee, and the earliest winner, is probably *Camelot* (1967). I say "probably" due to the previously noted uncertainty of the King Arthur timeline, but our sixth-century guess slots him into second place behind *Gladiator*.

On the other end of the spectrum, the dystopian film *Logan's Run* (1976) takes place in 2274. It's the latest nominee in this category, just barely beating *Star Trek* (1979), which is set in 2271.

These two are exceptions to the rule, because Production Design much more frequently looks to the past. Since 1967, only two Best Production Design winners have been set in the present day: *Heaven Can Wait* (1978) and *La La Land* (2016). Remarkably, *Heaven Can Wait* held the title of most recent Best Production Design winner all the way until *La La Land* took the crown, as no other winner in this category took place between 1978 and the present.

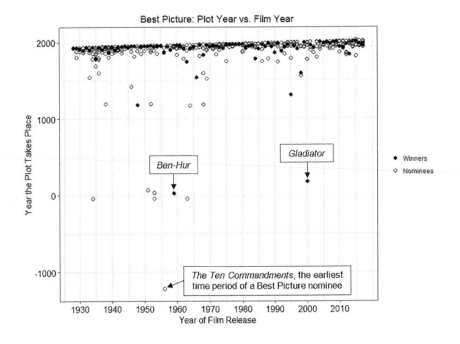

For the Best Picture graph, that point at the very bottom is *The Ten Commandments* (1956), the earliest Best Picture nominee by over a millennium. People do not agree on whether there truly was a historical Exodus, and even if there was one, scholars differ on when it took place. One commonly cited era is the thirteenth century BCE, so that's what I used. Even if that's off in either direction by a few centuries, Cecil B. DeMille's epic is still solidly in first place for the oldest contender in this category.

Only three others take place before the year 0: Shakespeare's *Julius Caesar* (1953) and both versions of *Cleopatra* (1934, 1963). The earliest Best Picture winner, *Ben-Hur* (1959), is also set in Roman times. We can make the graph more readable by removing films that take place before 1700:

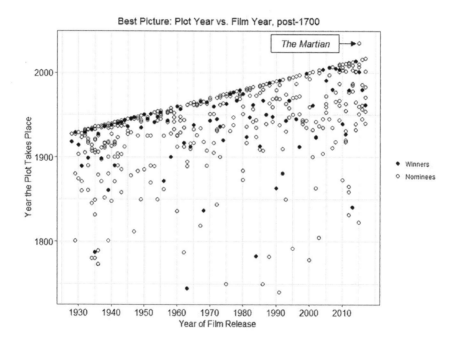

The first thing that stands out is the point in the top right. The latest Best Picture nominee with an exact year associated with it is *The Martian* (2015), as author Andy Weir confirmed that his Mars survival story is set in 2035.

The latest Best Picture winner, somewhat by default, is *Birdman*

(2014). When a nominee does not specify a year but is clearly intended to be set in the present day, I took the year of the film's release to be the year the story takes place. Many movies transpire in the present, explaining why the chart shows an upward slant over time. Therefore, *Birdman*'s record will almost assuredly be broken repeatedly as newer Best Picture winners are set in the present or future.

● ● ●

Historically, the most common winners for both Best Picture and Best Production Design are dramas that are set in New York and take place in the past. *The Great Gatsby* (2013), which claimed Best Production Design, even manages to check all three of those boxes.

But does this mean we've cracked the code for winning a Production Design Oscar? Not quite. These features – drama, New York, historical – also describe a whole lot of non-winning films, and even more un-nominated ones. In all likelihood, the best way to win this Oscar is to simply to design with aplomb whichever time and place the story calls for. Wherever and whenever that is, transport us there and then.

Chapter 8. Best Cinematography

Do career-achievement Oscars exist?

OFFICER:
Sir, allow us to pledge you the most
glorious victory of your career.

CRASSUS:
I'm not after glory! I'm after Spartacus.
And, gentlemen, I mean to have him.
However, this campaign is not alone to
kill Spartacus. It is to kill the legend
of Spartacus.

–*Spartacus* (1960): Nominated for 6 Oscars; won 4 (Best Supporting Actor, Best Art Direction: Color, Best Cinematography: Color, Best Costume Design: Color)

IN VICTOR Fleming's epic retelling of the martyrdom of *Joan of Arc* (1948), the title character played by Ingrid Bergman says of her impending demise, "No, the pain won't be little – but it will end."

She could just as easily have been foreshadowing the Oscar histories of her film's cinematographers, though no literal stake was involved and the figurative stakes were far lower. *Joan of Arc*'s cinematographers were Winton Hoch, William Skall, and Joseph Valentine. Hoch, a first-time nominee, would wind up as the only cinematographer in history to go three-for-three in his Oscar career, adding on wins for *She Wore a Yellow Ribbon* (1949) and *The Quiet Man* (1952).

But Skall and Valentine were veterans, and unlucky ones at that. This was Skall's eighth nomination, and he had no wins to show for it. Same goes for Valentine, who was on his fifth trip to the Oscars. Add up the three contenders, and they had a career 14 nominations and no wins before *Joan of Arc*'s name was pulled from the cinematography envelope.

Did this lack of prior success have anything to do with their win? Surely, voters might have been tempted to finally put a trophy on the shelf for their colleagues who had long produced high-quality photography. This was the same year that Skall and Valentine did cinematography for Alfred Hitchcock's tracking-shot thriller *Rope*, which received no Oscar nominations but, frankly, should have. Skall even notched three nominations in 1942 alone, something no other cinematographer has ever achieved in a single year, yet lost all three (for *Arabian Nights*, *Reap the Wild Wind*, and *To the Shores of Tripoli*) to *The Black Swan*.

We can't get inside the heads of 1948 Oscar voters, but we can use math to answer the broader question: Does the Academy in general give a boost to nominees who have not yet won? That is, in Best Cinematography as well as other competitive (i.e., non-honorary) categories, do the Oscars reward career-achievement? Or, do they judge each nomination on its own merits, independent of an artist's work on previous films?

To answer this question, it will be helpful to use categories without too many shared nominations. For example, in 2007, legendary Disney composers Alan Menken and Stephen Schwartz teamed up to write three Oscar-nominated songs for *Enchanted*: "Happy Working Song," "So Close," and "That's How You Know." Incredibly, all three lost to "Falling Slowly" from *Once*. In all likelihood, *Enchanted* fans split their votes, leading the Academy to change its rules and allow only two song nominations per film going forward.

This trio of nominations were Menken's 16th, 17th, and 18th, including Best Score wins for *The Little Mermaid* (1989), *Beauty and the Beast* (1991), *Aladdin* (1992), and *Pocahontas* (1995). He also won Best Song for a number from each of those films: "Under the Sea," "Beauty and the Beast," "A Whole New World," and "Colors of the Wind."

Schwartz, by comparison, was practically a novice. He had "only" five Oscar nominations prior to *Enchanted*, sharing Menken's two *Pocahontas* wins along with a solo trophy for "When You Believe" from *The Prince of Egypt* (1998). How do we judge those three *Enchanted* nominations? Do they count as career nominations #6, #7, and #8, or are they #16, #17, and #18?

So, we'll stick to the categories in which fewer than 10% of their nominations are shared among multiple individuals:[1]

---

1 In the graph, I left off the seven awards that go to an entire film, as opposed to individual components of a film.

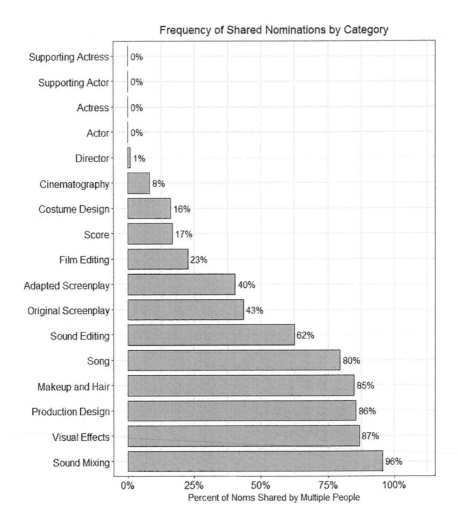

It turns out that the four acting categories (which by definition cannot be shared), Best Director, and Best Cinematography fit the bill. Best Director has only four shared nominations:

- *West Side Story* (1961), directed by Robert Wise and Jerome Robbins – Winner

- *Heaven Can Wait* (1978), directed by Warren Beatty and Buck Henry

- *No Country for Old Men* (2007), directed by Joel and Ethan Coen – Winner

- *True Grit* (2010), directed by Joel and Ethan Coen

Best Cinematography, at 8% shared, isn't quite as clean, but we're making a tradeoff: In exchange for allowing a few more shared nominations to mess with the data, we're getting another category's worth of nominees to examine, and more data tends to make conclusions more accurate, all else being equal.

Let's use those six categories and dive right in, first by examining how often a nominee enters an Oscar night with n prior nominations:

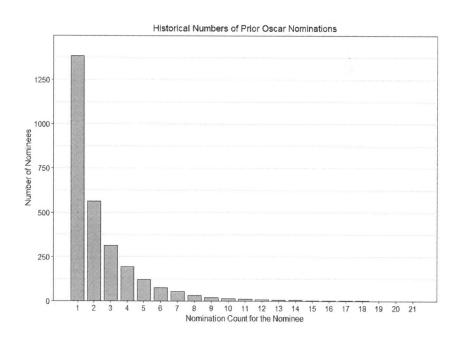

Everything to the right of 14 nominations represents just four individuals:[2]

- Meryl Streep (actress): 21 nominations, with wins for *Kramer vs. Kramer* (1979), *Sophie's Choice* (1982), and *The Iron Lady* (2011)

- Leon Shamroy (cinematographer): 18 nominations, with wins for *The Black Swan* (1942), *Wilson* (1944), *Leave Her to Heaven* (1945), and *Cleopatra* (1963)

- Charles Lang (cinematographer): 18 nominations, with a win for *A Farewell to Arms* (1932)

- Robert Surtees (cinematographer): 16 nominations, with wins for *King Solomon's Mines* (1950), *The Bad and the Beautiful* (1952), and *Ben-Hur* (1959)

Unsurprisingly, the chart shows more first-time nominees than second-time nominees, more seconds than thirds, and so on. But to see which count brings the most luck, let's look at this same data in a different way, examining the percentage of nominees at each career count who win their category:

---

[2] Across all categories, there are 52 individuals with 15+ Oscar nominations, led by Walt Disney with 59 and composer John Williams with 51.

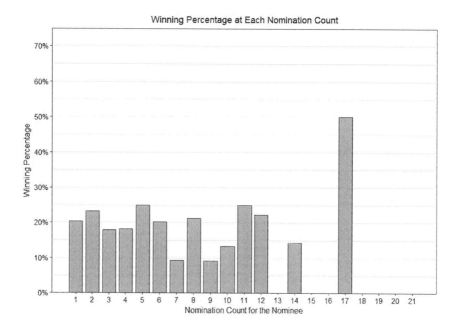

Technically, the most successful nomination count is #17, at 50%. But that represents just two wins out of four nominations: Shamroy's win for *Cleopatra* and Streep's for *Iron Lady*, with the losses for Shamroy's work in *The Cardinal* (1963) and Lang's cinematography for *Bob & Carol & Ted & Alice* (1969) – I count both Shamroy works as #17, since they occurred in the same year, and he had 15 nominations prior to that year.

Two out of four is hardly enough data to prove that a 17-time nominee is better off than a 16- or 18-time contender. But those bars on the left side of the graph represent a far greater number of nominees throughout history. With the exception of the strangely low success rate for seventh- and ninth-timers (so much for lucky number seven), which I'm guessing is just a fluke, none of the bars are particularly low or high.

It's worth noting that first-timers come in at a 20% success rate, the same as the overall average of 20% (one out of five nominees wins each category). So, there is neither beginner's luck nor a veteran's advantage at the Oscars.

But this doesn't quite answer the career-achievement question. Typically, when people talk about career-achievement Oscars, they're referring to wins for people who have enjoyed lengthy careers but have thus far failed to win. So let's look at the winning percentages for nominees who have and have not won yet, grouped by how many prior nominations they had.

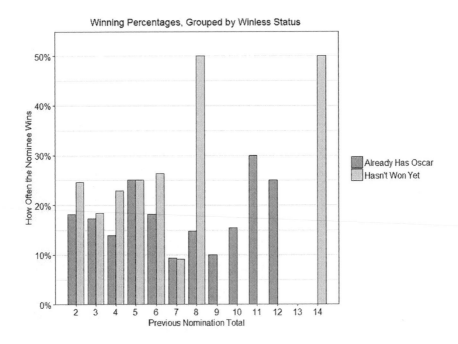

The chart begins on a nominee's second career nomination. On his or her first nomination, it's impossible for the nominee to be a previous winner, so we can't do a comparison between previous winners and still-winless nominees. The chart goes as far as the 14th nomination. Once we reach the 15th nomination or higher, all nominees in these categories have already secured at least one win.

The results from nominations #9 to #14 reflect very few actual cases. Only cinematographers Roger Deakins and George Folsey have made it that far — all the way to their ninth nomination or more — without having already notched a win. So, we shouldn't draw too many conclusions from a sample size of two cinematographers. Even nomination #8 is not based on a huge sample, as that light grey bar at 50% only means that three out of six winless nominees won the Oscar on their eighth try.

But nominations #2 to #7 tell us a story. On average, the light grey bars (nominees who haven't won an Oscar yet) are higher than the dark grey bars (nominees who have already won an Oscar). This is quite notable. All else being equal, we would actually expect the nominees with previous wins to outperform the perennial losers. After all, previous wins should indicate greater skill in the craft, be it acting, directing, or cinematography.

But something else is going on. This data suggests the existence of career-achievement awards, representing the philosophy that the Oscars should reward all-time greats who have yet to receive a trophy.

Overall, for nominees between their 2nd and 14th nomination, where we have data on both prior winners and still-winless contenders, the average prior winner has a 17.0% success rate. The average still-winless individual sits at 22.9%. That's a 5.9% boost for Oscar hopefuls still waiting for their inaugural trophy.

Whether or not this is a habit the Oscars "should" be engaged in is up to you, dear reader. On the one hand, I can't help but feel some sympathy for the extraordinarily deserving Hollywood legends who never came away a winner. The holders of the dubious record of most nominations without a win in each of the categories examined above are as follows:

Best Director[3]

- Clarence Brown (0/6): *Anna Christie* (1930), *Romance* (1930), *A Free Soul* (1931), *The Human Comedy* (1943), *National Velvet* (1944), *The Yearling* (1946)

Best Actor/Supporting Actor

- Peter O'Toole (0/8): *Lawrence of Arabia* (1962), *Becket* (1964), *The Lion in Winter* (1968), *Goodbye, Mr. Chips* (1969), *The Ruling Class* (1972), *The Stunt Man* (1980), *My Favorite Year* (1982), *Venus* (2006)

Best Actress/Supporting Actress

- Deborah Kerr (0/6): *Edward, My Son* (1949), *From Here to Eternity* (1953), *The King and I* (1956), *Heaven Knows, Mr. Allison* (1957), *Separate Tables* (1958), *The Sundowners* (1960)

- Glenn Close (0/6): *The World According to Garp* (1982), *The Big Chill* (1983), *The Natural* (1984), *Fatal Attraction* (1987), *Dangerous Liaisons* (1988), *Albert Nobbs* (2011)

- Thelma Ritter (0/6): *All About Eve* (1950), *The Mating Season* (1951), *With a Song in My Heart* (1952), *Pickup on South Street* (1953), *Pillow Talk* (1959), *Birdman of Alcatraz* (1962)

---

3 Some sources count Clarence Brown's nominations for two 1930 movies as a single nomination. In that case, Brown would be 0/5 instead of 0/6, and would tie with three other directors who also went 0/5: King Vidor, Alfred Hitchcock, and Robert Altman.

Best Cinematography

- George Folsey (0/14): *Reunion in Vienna* (1933), *Operator 13* (1934), *The Gorgeous Hussy* (1936), *Lady of the Tropics* (1939),[4] *Thousands Cheer* (1943), *The White Cliffs of Dover* (1944), *Meet Me in St. Louis* (1944), *The Green Years* (1946), *Green Dolphin Street* (1947), *Million Dollar Mermaid* (1952), *All the Brothers Were Valiant* (1953), *Executive Suite* (1954), *Seven Brides for Seven Brothers* (1954), *The Balcony* (1963)

When you think of some of this incredible work, such as Peter O'Toole's acting in *Lawrence of Arabia* and George Folsey's cinematography for *Meet Me in St. Louis*, you can easily understand the temptation to retroactively honor these artists – if not for that specific film, then at least for some future nomination in recognition of all they've done for cinema. Some critics asserted that the Academy did precisely this when it awarded Martin Scorsese Best Director for *The Departed* (2006) after bypassing him for *Taxi Driver* (1976), *Raging Bull* (1980), and *Goodfellas* (1990). As a compromise, the Academy occasionally hands out literal, noncompetitive career-achievement awards, including one to O'Toole in 2003.

On the other hand, each year brings a whole new crop of deserving contenders. So, if the Academy opts to give a competitive award to a career-achievement candidate instead of choosing the best work from that year's films, that risks creating a whole new set of historical injustices that will need to be rectified by future voters, and the cycle never ends. What's more, if the Academy considers the Oscars

---

4 Folsey's nomination for *Lady of the Tropics* is considered to be unofficial, since that year the Academy published both an initial list of potential contenders (which included Folsey) and a final list of nominees (which left him off).

to be a historical record of the best work from each season of movies, it somewhat distorts that record by using criteria from a nominee's previous work. There is not a right or wrong answer here.

There is, however, a right and wrong answer on whether or not the Academy currently gives a boost to winless nominees. Thanks to math, we now know that there's a 5.9% bump for entering Oscar night without a win to a nominee's name. And sometimes, that extra 5.9% is all it takes to finally put that name in the win column.

Chapter 9. Best Original Score

Do the Grammys and Tonys agree with the Oscars?

> **MOZART:**
> In a play, if more than one person speaks at the same time, it's just noise. No one can understand a word. But with opera, with music — with music you can have twenty individuals all talking at the same time, and it's not noise: It's a perfect harmony.
>
> –*Amadeus* (1984): Nominated for 11 Oscars; won 8 (Best Picture, Best Director, Best Actor, Best Adapted Screenplay, Best Art Direction, Best Costume Design, Best Makeup, Best Sound)

WHEN *An American in Paris* danced into theaters in 1951, it made history for having the guts to conclude a movie with 17 straight minutes of dialogue-free ballet, which cost $450,000 to produce.

One year later, at the Oscars held in 1952, the movie made more history, becoming the only film to win this combination of awards: Best Picture, Best Original Screenplay, Best Production Design, Best

Cinematography, Best Costume Design, and an award for scoring.

And then, 63 years later, it made even more history at the 2015 Tony Awards. The Broadway stage version of *An American in Paris* became the first (and to date, only) production to win Best Musical Score at the Oscars and later be nominated for its theatrical version for Best Musical at the Tony Awards (losing to *Fun Home*).

Wait a minute – Best Musical Score at the Oscars? Is that even a category? The answer is "depends on the year." The Academy introduced Best Original Score in 1934, delivering the category's first trophy to *One Night of Love*, a musical about an American opera singer in Italy. Many of the earliest winners in the new category were also musicals: *One Hundred Men and a Girl* (1937), *Alexander's Ragtime Band* (1938), *The Wizard of Oz* (1939), *Pinocchio* (1940), and *Tin Pan Alley* (1940)[1] all won Oscars for either their main scores or their underscores.

At that point, the Academy decided it was time for musicals to get their own Best Score category. Though the name changed frequently (for simplicity's sake, I will use the name Best Musical Score throughout this chapter), the Academy handed out a scoring award solely for musicals in 41 different years, with the last one going to Prince's *Purple Rain* (1984).

In 2000, the Academy added a Best Original Musical category to the program, but specified that it would only be handed out if enough high-quality musicals came out in a single year. It's not clear how many is enough, and thus far not a single year has been deemed worthy, so the category has remained dormant.

While there is only one Best Score award presently, history pro-

---

[1] Both *Pinocchio* and *Tin Pan Alley* won Oscars for scoring in 1940 since that was one of the years with two awards: Best Original Score went to *Pinocchio* for its main music, while Best Scoring went to *Tin Pan Alley* for its underscore, the quieter music performed underneath the on-screen dialogue or action.

vides us with both musical and nonmusical categories, allowing for interesting comparisons with the Oscars' cousins in the awards business: the Tonys and the Grammys. This chapter will compare the Oscar for Best Musical Score to the Tonys, and the Oscar for Best Original Score (the category not specific to musicals) to the Grammys.

Oscars vs. Tonys

Though *An American in Paris* stands alone in traveling from Hollywood to Broadway at awards shows' top musical categories, a number of musicals have gone the other direction. The following shows all first won Best Musical at the Tonys before winning Best Musical Score at the Oscars:

- *The King and I*: 1952 Tonys, 1956 Oscars
- *The Music Man*: 1958 Tonys, 1962 Oscars
- *My Fair Lady*: 1957 Tonys, 1964 Oscars
- *The Sound of Music*: 1960 Tonys, 1965 Oscars
- *A Funny Thing Happened on the Way to the Forum*: 1963 Tonys, 1966 Oscars
- *Hello, Dolly!*: 1964 Tonys, 1969 Oscars
- *Fiddler on the Roof*: 1965 Tonys, 1971 Oscars
- *Cabaret*: 1967 Tonys, 1972 Oscars
- *A Little Night Music*: 1973 Tonys, 1977 Oscars

These nine musicals won at both the Oscars and the Tonys, and all nine first premiered on Broadway and later won accolades for their film versions. As it turns out, this trend holds true across a wider sample: Musicals are more likely to journey from east to west than west to east. When we zoom out to include nominees and not just winners, we can compile a chart of all 36 musicals nominated for either Best Musical or Best Original Score at the Tonys plus Best Musical Score at the Oscars, arranged by how long the show took to cross over from Broadway to Hollywood or vice versa:

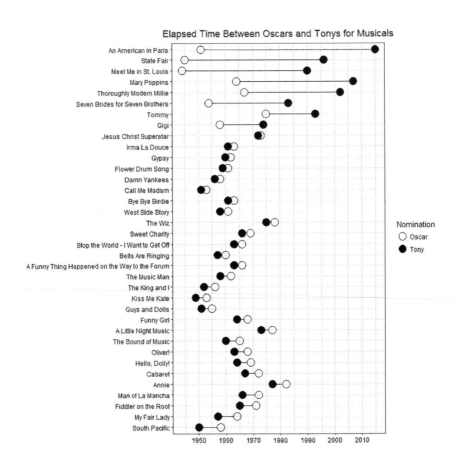

Two takeaways jump out from this data. First, the black dots (representing the Tonys) are usually to the left of the white dots (representing the Oscars), meaning New York first and L.A. second. To be precise, 28 of these 36 productions began on the Great White Way before conquering the silver screen.

Second, when the musical started at the Tonys, the gaps in time are all pretty short. The average elapsed time of the 28 musicals that originated in New York is just 3.8 years before landing in Hollywood. The longest gap between stage year and film year belongs to Rodgers and Hammerstein's *South Pacific* (1958), which took eight years to travel to a California studio. Andrew Lloyd Webber's *Jesus Christ Superstar* (1973) set the all-time speed record, getting to Hollywood just one year after taking the stage at the Tony Awards.

At the top of the chart, the musicals moving in the other direction have much lengthier elapsed times between screen and stage. In fact, the *shortest* of those gaps, the 16 years it took for *Gigi* (1958) to get from the City of Angels to the Big Apple, is still wider than the *longest* elapsed time from NYC to L.A., *South Pacific*'s 8 years.

What this means for theater fans: If you're hoping that your favorite Tony nominee from years ago will appear on an Oscar telecast near you, you might be out of luck. History suggests that, after eight years or so, it's fairly unlikely to see Hollywood take the bait. But on the flip side, people who fondly recall a movie musical from childhood still have hope that the show will one day grace the Tony stage, even decades later.

There are a few exceptions from outside this dataset. *Chicago* (2002), the only musical to win Best Picture in the last half-century, is not included because there was no Best Musical Score category in 2002. *Chicago* didn't land a Hollywood production until 27 years after the original Broadway production premiered, but even in that case, the movie version was likely spurred more by the 1996 Broadway revival than by Bob Fosse's 1975 original, so this anecdote arguably counts as further proof that Hollywood is reluctant to produce musicals based on decades-old stage shows.

In the years when the Academy did provide a standalone category for musicals, was there any correlation between its tastes and those of its Manhattan counterpart? Let's examine the data. Eighteen Best Musical winners at the Tonys have also been nominated in the equivalent category at the Oscars. In two years, a pair of Broadway champions actually competed against each other:

- In 1958, Tony winners *South Pacific* and *Damn Yankees* faced off at the Oscars, but both lost to *Gigi*. Sixteen years later, *Gigi* found its way to New York, winning Best Score but losing Best Musical.

- In 1972, a pair of back-to-back stage victors both traveled to California: *Man of La Mancha* and *Cabaret*, which won Best Musical at the Tonys in 1966 and 1967, respectively. When pitted face-to-face at the Oscars, *Cabaret* emerged triumphant.

In 16 Oscar seasons, at least one Best Musical Tony winner competed for the Best Musical Score Oscar. In 14 of those seasons, just one Tony winner was up for the award. In the other 2 of those seasons, the pairs of Tony winners listed above went head-to-head. I asked my computer to perform a simulation of those 16 seasons – in other words, to rerun each of them 10,000 times, randomly choosing a winner from among each year's nominees – to see how often we would expect the Tony winners to also win the Oscar, if the Oscars were completely random:

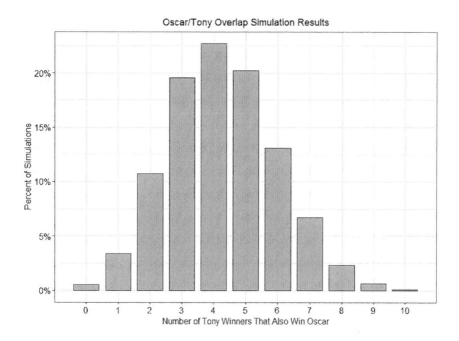

In practice, the correct answer is nine – specifically, the nine films listed earlier in the chapter that won both Tony and Oscar honors for musicals. Since only 0.8% of simulations had a result of nine or higher, it's safe to say that the high number of overlaps in real life compared to the simulation demonstrates that this result is more than a coincidence. The Tonys and the Oscars do prefer the same musicals more often than random chance would suggest.

Oscars vs. Grammys

How about the Tonys' and Oscars' friend in the music industry, the Grammys? That show also has an award for Best Score, titled Best Score Soundtrack for Visual Media. Let's see how well that award correlates with the Oscar for Best Original Score (the category that is not specific to musicals) throughout history.

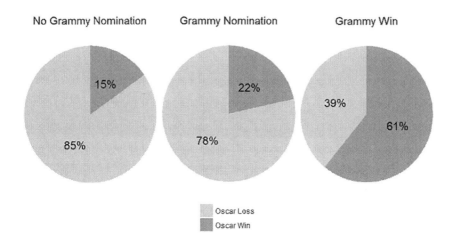

Movies that garner Grammy nominations are more likely to win Best Score at the Oscars than those without a Grammy nomination. And movies that win Grammys are way more likely to emerge triumphant at the Oscars. None of this proves cause and effect because these awards shows follow different calendars. Sometimes the Grammy honor predates that of the Oscars, and sometimes it's the other way around. But the data does demonstrate correlation between the two bodies' preferences.

Twenty-five films won both the Oscar and the Grammy for Best Score, though in 13 of those cases, the Oscars made it easier by having multiple awards for Best Score that year. That leaves 12 films in more recent years (many of which only had one Best Score category) that were the sole Best Score winners at both the Oscars and Grammys:

- *Out of Africa* (1985)
- *The Last Emperor* (1987)
- *Dances with Wolves* (1990)
- *Beauty and the Beast* (1991)

- *Aladdin* (1992)
- *Schindler's List* (1993)
- *Crouching Tiger, Hidden Dragon* (2000)
- *The Lord of the Rings: The Fellowship of the Ring* (2001)
- *The Lord of the Rings: The Return of the King* (2003)
- *Up* (2009)
- *The Grand Budapest Hotel* (2014)
- *La La Land* (2016)

The Grammys have a decent track record of at least nominating Oscar winners. Only nine sole Best Score winners at the Oscars have gone un-nominated by the music awards:

- *Chariots of Fire* (1981)
- *'Round Midnight* (1986)
- *The Milagro Beanfield War* (1988)
- *Frida* (2002)
- *Finding Neverland* (2004)
- *Brokeback Mountain* (2005)
- *Atonement* (2007)
- *Slumdog Millionaire* (2008)
- *The Social Network* (2010)

All of the datasets I use in this book will of course change over the years, decades, and (hopefully) centuries as the Oscars generate more data. What's unique about this chapter is that the data could theoretically change even for past years, not just future ones.

To choose an arbitrary example, MGM produced a delightful musical called *Anchors Aweigh* in 1945 about a couple of sailors (Gene Kelly and Frank Sinatra) on leave in Hollywood. The movie won Best Musical Score. Thus far, no Broadway producer has ever brought this tale to the Great White Way. If one day, 10 or 50 or 100 years from now, someone decides to do so, *Anchors Aweigh* could be added to the list of musicals nominated by both the Oscars and the Tonys, and could smash the record currently held by *An American in Paris* for longest elapsed time between nominations.

With that said, the Broadway producer would face a daunting task in recreating the film's most famous scene, when Gene Kelly dances with cartoon Jerry Mouse. Perhaps some things are better left to movie magic.

Chapter 10. Best Original Song

Music Theory at the Oscars

CONDUCTOR:

What are you going to sing, Ms. Lamont?

KATHY:

(behind the curtain) "Singing in the Rain."

LINA:

(on stage) "Singing in the Rain."

CONDUCTOR:

"Singing in the Rain." In what key?

KATHY:

(behind the curtain) A-flat.

LINA:

(on stage) A-flat.

CONDUCTOR:

In A-flat.

KATHY:

(behind the curtain) I'm singin' in the rain, just singin' in the rain. What a glorious feelin', I'm happy again.

—*Singin' in the Rain* (1952): Nominated for 2 Oscars; won 0

BEST ORIGINAL Song gets more airtime during the Oscar ceremony than any other category because in most years all nominees perform their songs on stage. Perhaps we sing along to the ones that became pop hits that year, but rarely do we consider the structure of the compositions written for the silver screen.

Let's analyze some of the winners of the Oscar for Best Original Song via the discipline of music theory. This chapter will give the talented composers behind cinema's greatest tunes their due for just how clever their music and lyrics are, perhaps even more clever than we realize at first blush.

Melody

Frank Sinatra sang arguably the cheeriest Best Song winner in "High Hopes" from *A Hole in the Head* (1959), the story of a man who comes to realize what's truly important in life. To emphasize just how high our hopes should be, composers Jimmy Van Heusen and Sammy Cahn were sure to emphasize the word "high" with high notes. Specifically, after all 12 mentions of the word "high" or "sky," the next note is always lower, coaxing the listener to experience the descent from that high place.

Other musicians incorporate bits of well-known songs to make audiences feel a certain way or put them in a certain mindset. Two notable examples from Oscar-winning songs are "The Last Time I

Saw Paris" from *Lady Be Good* (1941) and "Skyfall" from the 2012 film of the same name.

"The Last Time I Saw Paris" starts and ends with the opening bar of "La Marseillaise," the French national anthem. The song, along with its scene in the film, deals with nostalgia for prewar Paris before the German invasion.

In a similar vein, Adele cleverly slipped in the slow note progression B – C – C# – C at the end of the "Skyfall" chorus. Those are the exact same notes that open the iconic James Bond franchise theme song, so they immediately evoke a 007 mood in an audience looking forward to the latest episode of suave British spycraft.

Time Signature[1]

The vast majority of Oscar-winning songs, not to mention pop songs, are composed in 4/4 or some variation, meaning they generally have an even number of beats per measure. So, when a song does not have an even time signature, that represents a conscious decision by the composer, sometimes for a plot-driven reason. A good example is "Chim Chim Cher-ee" from *Mary Poppins* (1964), written in 3/4 to conveniently match the beat of children skipping, which is precisely how the song begins, with Bert, Jane, and Michael holding hands as they skip in the park.

Another Oscar-winning example is "You Light Up My Life" from the 1977 film of the same name. The composer uses 3/4 time to convey that the protagonist Laurie is both a gifted songwriter and a nervous auditioning singer. Laurie shows her talent by writing a full-orchestra piece outside the bounds of ordinary pop, but also demonstrates her anxiety

---

1   Time signature refers to how many beats are in a measure and what type of note counts for a single beat.

when she reminds the conductor that the song is in 3/4, a fact he already knows by looking at the music. Not only does she nail the audition in the movie, but the song in real life reached #1 on the *Billboard* Hot 100 for ten consecutive weeks, breaking the record previously held by Percy Faith's "Theme from *A Summer Place*" and the Beatles' "Hey Jude."

Genre

Ava DuVernay's *Selma* (2014) is set during the fight for civil rights in 1965 Alabama, but it is also meant to deliver a message about current efforts for equality under the law. Fittingly, when Common and John Legend wrote "Glory," they gave it twin genres: a gospel ballad for the chorus and bridge, and hip-hop for the verses. This brilliantly mirrors the message of the song, that "Selma is now for every man, woman, and child." The lyrics switch back and forth effortlessly between references to the 1960s civil rights battle and current campaigns for justice ("That's why we walk through Ferguson with our hands up"). In the same way, the song alternates between a musical style that would have been very familiar to Dr. King and his contemporaries, and one that resonates more with the audience watching the film. The music signifies to the listener that the struggle for justice is not a merely a relic of history; it continues to the present day.

In Damien Chazelle's *La La Land* (2016), Ryan Gosling's character Sebastian wants to save classical jazz from modern influences. And the movie itself not so subtly shares that goal, never more so than in the scene when Sebastian and Mia sing a duet of a Justin Hurwitz song, with lyrics by Benj Pasek and Justin Paul, called "City of Stars." It's slow, it's passionate, it's nostalgic, and most importantly, it never breaks from the traditional jazz rhythm: a long first beat followed by a short second beat, over and over. The song serves the purpose not only of the characters singing it but also of the film itself.

Key

Elton John and Tim Rice penned one of cinema's most powerful love songs in "Can You Feel the Love Tonight" for *The Lion King* (1994). As Timon and Pumbaa lament losing their friend Simba to something as silly as love, they sing in F major while a montage of Simba and Nala's budding romance begins. Then, suddenly, the key climbs to G major and the chorus returns to "Can You Feel the Love Tonight" with greater bravado, signaling something dramatic is about to happen. At that moment, Nala kisses Simba for the first time (or, to be precise, she holds his face with her paw and licks his cheek, which is clearly Disney's way of implying that these lions are kissing).

That type of key change, moving up one or more notes before a dramatic final chorus, is by far the most standard one in Western music. Examples include Whitney Houston's "One Moment in Time," Jon Bon Jovi's "Livin' on a Prayer," and Taylor Swift's "Love Story." But a year after *The Lion King*, another Disney film used a more unique key change to tell a story, putting one at the beginning of a song.

"Colors of the Wind" from *Pocahontas* (1995) begins in D minor, a somber tone to reflect the racism ("You think I'm an ignorant savage") and ignorance ("How can there be so much that you don't know?") of the European invaders. Then, the animators draw a huge gust of wind colored by leaves that fly past Pocahontas, and the key abruptly changes to a sprightly D major. From there, the singer begins to poetically teach John Smith about the beauty and spirit of nature.

Tempo

Judy Garland performed what is, in my opinion, the best use of tempo in an Academy Award song winner. In *The Harvey Girls* (1946), she sings "On the Atchison, Topeka, and the Santa Fe" to celebrate the impending departure of the titular train line. Songwriting duo Harry

Warren and Johnny Mercer cleverly wrote the song to mirror the sound of a train leaving a station. It starts off slowly, with the swingtime beat of a train just starting to roll. Garland sings the word "whistle" just as a train whistles in the background. The song gains speed, eventually culminating in a full chorus of singers dancing alongside a moving train, which itself serves as a unique form of percussion as the number drives towards its conclusion.

In the 1980s, a pair of Oscar-winning songs from iconic dance films took full advantage of increasing beats per minute. "Flashdance… What a Feeling" from *Flashdance* (1983) and "(I've Had) The Time of My Life" from *Dirty Dancing* (1987) both begin leisurely, then change midway to a significantly quicker pace. This allows the choreographers to show off their dancers' versatility, dazzling the *Flashdance* judges and the *Dirty Dancing* resort audience with both luscious slow moves and snappy fast ones all in one famous number.

Instrumentation

The most masterful instrument selection in an Oscar-winning song was concocted by Alan Menken and Howard Ashman for "Under the Sea" from *The Little Mermaid* (1989). Ashman's lyrics promise "We got a hot crustacean band," and Menken's tune doesn't disappoint. Every sea creature is assigned an instrument to play, and the instruments themselves are represented on-screen by other marine life that resembles the instruments we hear. When Sebastian and a blue lobster "play" some clams, we hear marimbas, plus one hit of a cymbal. When "the sturgeon and the ray – they get the urge and start to play," their plants sound like a trumpet and maracas. We are told that "the newt play the flute, the carp play the harp, the plaice play the bass," though the bass is really two octopi with their legs tied together to resemble the instrument's strings. Sebastian fails to convince Ariel to abandon her dream to "explore that shore up above," but he succeeds in conducting the coolest band in oceanic history.

The newt isn't the only one who has mastered the flute. "My Heart Will Go On" from *Titanic* (1997) memorably opens with a wistful flute, setting up a tune that looks back on a lost love. The entire score of the movie, but particularly this song and choice of instrument, harkens back to traditional Irish flute music, some of which hails from the town of Belfast where the doomed ship was constructed.

Chord Progression

"Baby, It's Cold Outside" is now more famous as a mainstay of Christmas radio than as an Academy Award winner, but Frank Loesser (author of Broadway musicals *Guys and Dolls* and *How to Succeed in Business Without Really Trying*) won an Oscar for the song's inclusion in the romantic comedy *Neptune's Daughter* (1949). With the benefit of hindsight and improved social mores, the song has become controversial – is one singer attempting to coerce the other, or is this just a coy form of flirting? Like the lyrics themselves, the music ambiguously provides evidence for both arguments. On the one hand, the two singers are on different tunes most of the time and come in on different beats of most measures, implying that they're not on the same page in this liaison. But on a subtler level, the pair are nearly always singing the same chord in their ping-pong duet. Consider the opening lines:

It doesn't take familiarity with reading sheet music to see that the two singers croon different melodies. But the notes are actually the same: F – D – C – B♭ – F for the first singer and F – D – C – B♭ – D – B♭ – B♭ for the second. Both put the emphasis on B♭, D, and F, forming a B♭ major chord to open their parallel songs, a pattern that repeats often throughout the melody. The next lines each form C minor chords, and so on. So perhaps Loesser is slyly alluding to more agreement between the pair than first meets the ear. The script writer of *Neptune's Daughter* might agree, as both pairs of people who perform the song wind up falling in love.

A couple decades later, *The Thomas Crown Affair* (1968) employed a brilliant chord progression to tell a story, perhaps intentionally, perhaps not. The film contains a rather circular plot, with Steve McQueen and Faye Dunaway going round and round as both cop-and-robber and lovers. The featured song which won an Oscar, "Windmills of Your Mind," drives home that theme, opening with lyrics that contain nearly every imaginable circle-related term:

> Round like a circle in a spiral
> Like a wheel within a wheel
> Never ending or beginning
> On an ever-spinning reel
> Like a snowball down a mountain

Or a carnival balloon
Like a carousel that's turning
Running rings around the moon
Like a clock whose hands are sweeping
Past the minutes of its face
And the world is like an apple
Whirling silently in space
Like the circles that you find
In the windmills of your mind

But the music works in circular references on an even deeper level. For this analysis, it's necessary to get a bit into the weeds on music theory. Western music has 12 pitches (C, C#, D, E♭, E, F, F#, G, A♭, A, B♭, B). A "fifth" is the fifth note from a given starting pitch in any major or minor key. For example, in the key of C major, the first five notes are C, D, E, F, and G, so the "fifth" in that key is G.

We can continue to find fifths in any key. In the key of G, the fifth is D. In the key of D, the fifth is A. And so on. We will eventually hit all 12 pitches before arriving back at C, where we started. If we arrange these in a circle, commonly known as the "circle of fifths," it looks like this:

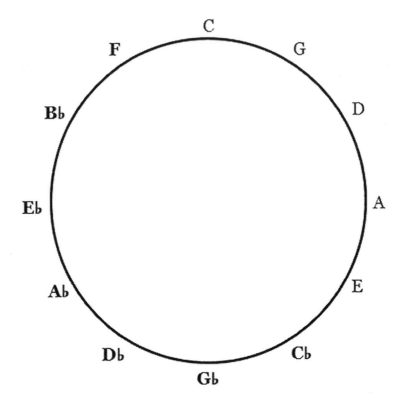

Most songs use chords that are in the key the song is written in. For example, if a song is written in the key of C, we are much more likely to hear a G chord than a G# chord, because G is in the key of C but G# is not. The song "Windmills of Your Mind" is written in the key of E♭ minor, so in the circle of fifths I have bolded the notes in that key (and referred to B as C♭ since that's how it's commonly referred to in this key).

Now, let's look at the chords that make up "Windmills of Your Mind" (you'll see why in a minute). In order, they are:

> E♭m, B♭7, E♭m, E♭7, A♭m7, D♭7, G♭maj7,
> C♭maj7, Fm7♭5, B♭7, Adim, B♭7, E♭m

This may look confusing for those who are unfamiliar with chord notation. These chords all start with a base note (for instance, the first chord starts with E♭). Then, they're followed by other symbols such as "m" (meaning it's a minor chord) or "7" (meaning an extra note, the seventh note in that scale, is added on). To simplify this, let's again write out the chords of "Windmills of Your Mind," but this time only focusing on the base notes in the key of E♭ minor:

E♭, B♭, E♭, E♭, A♭, D♭, G♭, C♭, F, B♭, E♭[2], B♭, E♭

Like the preschool game of connect-the-dots, let's connect these 13 notes in order on the circle on fifths:

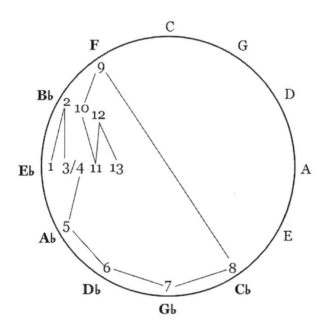

---

2 Musicians will notice that I interpret the base note of Adim (shorthand for A Diminished) as E♭. This isn't technically correct, but if you look at the three notes in Adim (A, C, E♭), the only one that exists in the key of E♭ minor is in fact E♭. If you'll permit me this liberty, the analysis that follows holds true.

Remarkably, every two adjacent chords are immediately next to each other within the subset of notes in the key of E♭ minor. There is never a chord transition that moves from, say, D♭ to F, because that's not a perfect fifth. The chord progression always moves in a circle, the circle of fifths, echoing "the circles that you find in the windmills of your mind."

Runtime

Gold Diggers of 1935 centers on a luxury hotel, where the wealthy but penny-pinching Mrs. Prentiss is producing a show that will cost her (and the real-life Warner Brothers studio) a small fortune. For the plot to work, it wouldn't be believable to make a three-minute song-and-dance number and claim that Mrs. Prentiss had to pay an enormous sum for it. So, Busby Berkeley directed an epic, charming, 13-minute-and-58-second dance extravaganza set to the upbeat tune of "Lullaby of Broadway." This number could have also been included in the section on genre, because it cleverly blends elements of musical stylings from both lullabies and show tunes, just as the title promises, to tell the story of a women who spends her days sleeping and her nights enjoying life on Broadway. Though it was the second winner of Best Original Song (following "The Continental" from *The Gay Divorcee* in 1934), "Lullaby of Broadway" still holds the record for the longest runtime of any winner in this category.

• • •

This is just a sampling of the clever tactics composers have used over the years. I somehow wrote a whole chapter on Best Original Song without analyzing "Over the Rainbow" from *The Wizard of Oz* (1939), or "White Christmas" from *Holiday Inn* (1942), or "Fame" from *Fame* (1980), or "Beauty and the Beast" from *Beauty and the Beast* (1991), or "A Whole New World" from *Aladdin* (1992), or "Lose Yourself" from *8 Mile* (2002), or "Let It Go" from *Frozen* (2013), or so many other impressive and memorable compositions.

These winning songs advance the plot, or develop a character, or set the mood, or sometimes just enhance the credits. So often we leave the theater humming a delightful tune. And you don't need any music theory to enjoy that.

Chapter 11. Best Sound Editing

Do big-budget films win more Oscars?

> PETER:
>
> Dollar and sixty cents. You had four dollars last night. How do you expect to get to New York at the rate you're going?
>
> ELLIE:
>
> That's none of your business.
>
> PETER:
>
> You're on a budget from now on.
>
> –*It Happened One Night* (1934): Nominated for 5 Oscars; won 5 (Best Picture, Best Director, Best Actor, Best Actress, Best Adapted Screenplay)

WHEN YOU watch the RMS *Titanic* crash into an iceberg in James Cameron's 1997 shipwreck epic, you're seeing a tragedy unfold in the Atlantic Ocean. But you're actually hearing ice break in Yellowstone National Park. Sound engineer Christopher Boyes recorded himself crushing ice in America's first park, and that's the sound that made it into the film.

Just like every other clever sound effect in the movie, the trick works perfectly. Boyes won an Oscar for Best Sound Editing for *Titanic* along with fellow sound editor Tom Bellfort. It was one of 11 Oscars won by the film that night, tying *Ben-Hur* (1959) for the most Oscars won by a single film, a figure later matched by *The Lord of the Rings: The Return of the King* (2003).

Unfortunately for producers, some sound effects cost far more than breaking ice. Even *Titanic*, having used that trick in Yellowstone, still spent its way to the title of most expensive Best Picture nominee of all time, adjusted for inflation. In modern times, movies with high-level sound creation tend to also have big budgets, both because producing the sounds themselves costs money and because the sorts of films that need impressive sound editing (action, adventure, etc.) often require those big budgets for other aspects of the movie, such as innovative visual effects.

Let's look at a graph of film budgets over time, both in the Best Picture and Best Sound Editing categories. I used Best Sound Editing data since 1988 because in some earlier years the category didn't exist at all and in others the category was honorary rather than competitive.

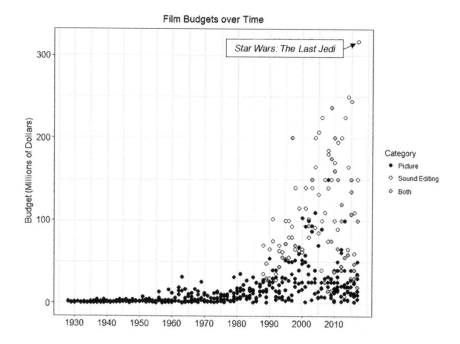

Clearly, movies have become a lot more expensive to make in recent years. Part of that increase is due to inflation, so let's look at the same graph adjusted to 2018 dollars:

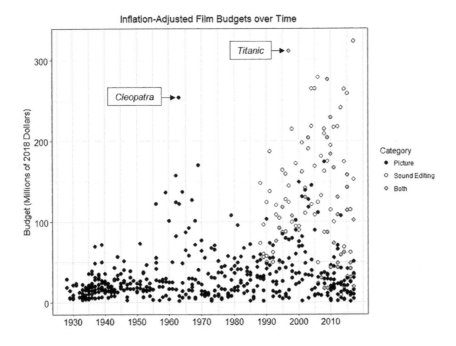

In both graphs – unadjusted and adjusted for inflation – the white dots (Best Sound Editing) are generally higher than the black dots (Best Picture). That is, Best Sound Editing nominees are usually more expensive to produce than Best Picture contenders. The fact that a disparity exists shouldn't be too surprising, given that the two categories have very little overlap. Indeed, only three films have won both – *Braveheart* (1995), *Titanic* (1997), and *The Hurt Locker* (2008) – which is the smallest number of overlapping winners between Best Picture and any other category, save for the six that honor entire films (Best Animated Feature, Best Documentary Feature, Best Foreign Film, and the three short film categories):

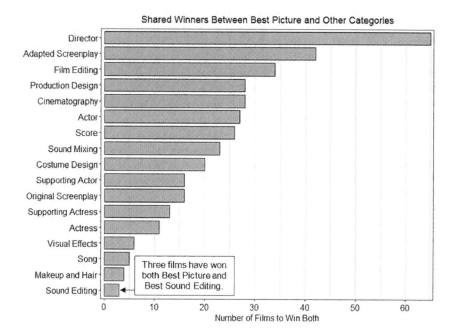

Because there's so little overlap between Best Picture and Best Sound Editing, we will analyze the relationship between budget and Oscar results for the two categories separately.

Does the budget help us predict the Oscars? If so, that would mean it's possible for studios to, in effect, buy themselves a higher chance of winning an award by pouring more money into a film. This chart shows how much the winners and nominees at the Oscars come in above or below the average budget of their competitors:

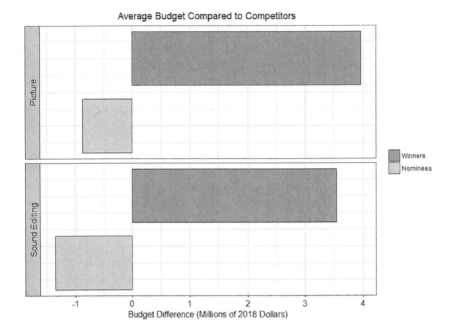

For both categories, Best Picture and Best Sound Editing, winners tend to spend about $5 million more to make their movies than those nominees who don't win. On the scale that movies are made on, tens or even hundreds of millions of dollars, this is not a huge difference, but it is enough to suggest that there might be some connection. Especially with Sound Editing, a relatively young category compared to Best Picture, it will be helpful to revisit this question in a few decades when we have more data. Then we can ascertain whether this is due to random noise or whether studios can spend their way to Oscar glory.

A higher budget can help not just with winning awards but also with earning revenue. On average, movies that cost more tend to sell more tickets, as seen on this graph comparing budget to box office:[1]

---

1 I am leaving the enormously lucrative *Gone with the Wind* off the graph since it's such an outlier that it would make the rest of the graph unreadable.

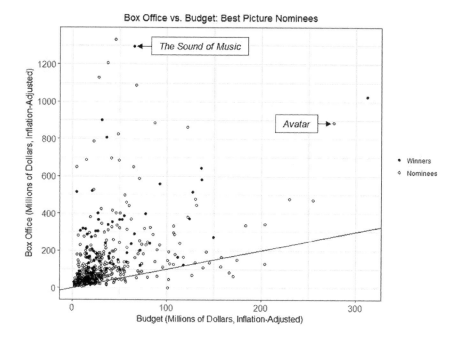

Points that lie along that diagonal line earned a profit of exactly $0, meaning they made as much as it cost to produce the film. Points above that line made money, while points below that line lost money. As the graph illustrates, the vast majority of movies turn a profit, a good outcome for the film industry.

But no matter how much a production company spends, there's still no replacing good old-fashioned filmmaking ingenuity, and that can't be bought. Just remember, even the most expensive Best Picture nominee of all time still needed a guy standing on ice to bring an epic soundtrack to our ears.

Chapter 12. Best Sound Mixing

How closely related are Sound Mixing and Sound Editing?

> HIGGINS:
> The majesty and grandeur of the English language is the greatest possession we have. The noblest thoughts that ever flowed through the hearts of men are contained in its extraordinary, imaginative, and musical mixtures of sounds. And that's what you've set yourself out to conquer, Eliza. And conquer it you will.
>
> –*My Fair Lady* (1964): Nominated for 12 Oscars; won 8 (Best Picture, Best Director, Best Actor, Best Musical Score, Best Art Direction: Color, Best Cinematography: Color, Best Costume Design: Color, Best Sound)

IN THE heyday of silent movie houses, live organists often accompanied films in real time. That's how important sound is to a movie – even before technology made it possible to prepackage audio and video, the motion picture industry was creatively delivering both to the audience.

If you haven't had the pleasure of watching a silent film yet, go watch one. Whether you opt for Harold Lloyd's daringly hilarious exploits in *Safety Last!* (1923), or the masterful recreation of World War I in *The Big Parade* (1925), or the timeless romance *Sunrise* (1927), or literally any Charlie Chaplin movie [try *City Lights* (1931), *Modern Times* (1936), and *The Circus* (1928), to name a few gems], silents are a delightful and awe-inspiring part of film history well worth your time.

Even without that experience, you can still appreciate the value of sound in scenes without talking. Consider the tragically beautiful score in the opening scene of *Up* (2009) or the mix of music and war sounds throughout *Platoon* (1986), and you can understand how sound makes all the difference even in the absence of dialogue, just as those live organists realized a century ago.

Once talkies arrived on the scene, sound became even more important. Sound editing and mixing made *King Kong* (1933) roar and *The Sound of Music* (1965) sing and *Saving Private Ryan* (1998) fight. Recognizing early on how vital this discipline is, the Academy introduced a Best Sound Recording category at the third ceremony. The inaugural award went to *The Big House* (1930) for firing up the sounds of a bloody prison riot in theaters nationwide.

That Best Sound Recording race was all well and good until 1963, when the Academy broke the category in two. Best Sound (the predecessor of Best Sound Mixing) went to *How the West Was Won* that year, while Best Sound Effects (Best Sound Editing in today's parlance) went to *It's a Mad, Mad, Mad, Mad World*.

Somewhere along the way, as the categories recombined and split once again and continually changed names, a whole lot of people lost track of which award was for which skill. To put it plainly, Sound

Editing as it is defined today is the art of creating sounds. Think of battle sounds in *The Incredibles* (2004), car chase noises in *The Dark Knight* (2008), and gunfire in *American Sniper* (2014). All three of those riveting audio tracks won Oscars for Best Sound Editing.

Sound Mixing is the art of blending sounds together. Listen to the crowd's jeers, warriors' grunts, and sword-fighting swishes melting together in *Gladiator* (2000). Recall the powerful voices singing in harmony in *Dreamgirls* (2006). Hear the intense jazz band keeping time in *Whiplash* (2014). Best Sound Mixing champions, one and all.

It is completely understandable that some audiences do not fully appreciate the difference between these two. If sound artists do their jobs correctly, the audio track sounds so authentic that you'd never realize many of those noises were created in a Hollywood studio independently from the on-screen action.

Plus, the category naming is a little confusing. (Personally, I'd be fine with reverting to one of Best Sound Editing's old names, Best Sound Effects.) Even Oscar devotees who follow the full awards calendar can get tripped up: Most other awards shows do not honor sound at all, and many of the ones that do (such as the BAFTAs) only have one award for sound, not two.

But surely, Academy voters are aware of the distinction, right? They work with sound experts on a regular basis and see all of the movies before filling out their ballots? In a perfect world, yes. But let's use data to learn just how well the Academy differentiates between these two skills.

As of the awards for 2017 films, the most recent Oscars before this book went to press, the Academy has bestowed both Sound Editing and Sound Mixing trophies for 38 awards seasons. In 16 of those years, the two awards went to the same film. Is that a lot or a little?

Luckily, we have a ready-made point of comparison. While math can never tell us the true quality of a movie's sound mixing or editing, we do have a decent proxy: the guilds representing audio professionals. The Cinema Audio Society (CAS) has handed out an award for achievement in sound mixing each year since 1993. The Motion Picture Sound Editors (MPSE) have honored elite sound editing since 1953.

In the 25 years since 1993, those two guilds have chosen the same winner 8.5 times, or 34%. Wait a minute – what's with the .5? In 2015, the MPSE announced a tie between *Mad Max* and *The Revenant*, the latter of which also won the CAS honor. We'll call that level of agreement 0.5. (Two decades earlier, *Braveheart* and *Crimson Tide* tied at the 1995 MPSE Awards, but neither of those films won the CAS race, which went to *Apollo 13*.) By comparison, the Academy chose the same movie for Best Sound Mixing and Editing in 13 of those 25 years. That's 52% alignment, ahead of the guilds' 34% agreement.

As this chart represents, that gap has widened in recent years. I have shaded in grey the years when the Oscars chose the same movie for both sound awards or when the CAS and MPSE agreed on the sound winner. As you can see, the Oscars' side of the table has a lot more grey:

Year	Oscar Sound Mixing	Oscar Sound Editing	CAS Sound Mixing	MPSE Sound Editing
2005	King Kong	King Kong	Walk the Line	War of the Worlds
2006	Dreamgirls	Letters from Iwo Jima	Dreamgirls	Letters from Iwo Jima
2007	The Bourne Ultimatum	The Bourne Ultimatum	No Country for Old Men	The Bourne Ultimatum
2008	Slumdog Millionaire	The Dark Knight	Slumdog Millionaire	The Dark Knight
2009	The Hurt Locker	The Hurt Locker	The Hurt Locker	Avatar
2010	Inception	Inception	True Grit	Inception
2011	Hugo	Hugo	Hugo	War Horse
2012	Les Miserables	Skyfall / Zero Dark Thirty	Les Miserables	Skyfall
2013	Gravity	Gravity	Gravity	Gravity
2014	Whiplash	American Sniper	Birdman	American Sniper
2015	Mad Max	Mad Max	The Revenant	The Revenant / Mad Max
2016	Hacksaw Ridge	Arrival	La La Land	Hacksaw Ridge
2017	Dunkirk	Dunkirk	Dunkirk	Blade Runner 2049

The Academy chooses the same winning movie for both sound categories much more frequently than the sound mixing guild and sound editing guild agree with one another, especially in recent years. In 2005, *King Kong* became the only movie ever to win both sound Oscars despite losing both sound guild awards. That marks the start of a rift between the Oscars and the sound guilds. In the 13 years since, the CAS and MPSE have only agreed with each other on 1.5 winners: the aforementioned *The Revenant* as well as *Gravity* (2013). That's 12%. In that same time frame, the Academy has presented both sound honors to the same movie 8 times, or 62%.

While only the members of the Sound Branch of the Academy choose the nominees in these categories, all 8,000+ Academy members have the chance to vote on the winners. Therefore, it's possible that some voters from outside the Sound Branch have a difficult time separating the two sound categories or that they don't see all of the nominees, and hence are more likely to vote for the same film twice.

Like this entire book, this chapter is dealt the curse of small sample sizes. Without more years' worth of data, it's difficult to say with certainty that the high level of sound mixing/sound editing agreement at the Oscars is not just a momentary blip. But given the data that is available, you can rest assured that if you sometimes get confused between Sound Mixing and Sound Editing while watching the Oscars, you may not be alone. You might have some company among a few members of the Academy.

Chapter 13. Short Film Categories

Does runtime affect the Oscars?

> **YUSUF:**
> Brain function in the dream will be about twenty times normal. When you enter a dream within that dream the effect is compounded. It's three dreams. That's ten hours, times twenty...
>
> **EAMES:**
> I'm sorry, math was never my strong subject. How much time is that?
>
> –*Inception* (2010): Nominated for 8 Oscars; won 4 (Best Cinematography, Best Visual Effects, Best Sound Mixing, Best Sound Editing)

IN THE 1971 animated short *The Crunch Bird*, a woman is struggling to choose a birthday gift for her husband. Eventually, she walks into a pet store and becomes enthralled with a "crunch bird," a unique creature that violently tears to shreds any item said after its name. The pet store owner demonstrates this with "crunch bird, the chair," and soon the chair is nothing but a pile of wood. Why the pet store even carries such a destructive bird is beyond me, especially since the pet store owner attempts to dissuade the woman from buying it. Nevertheless, she brings the dangerous pet home to her husband.

Rather than express gratitude for the birthday present, the husband is livid that his wife spent money on a pet. Unaware of the bloodshed he's inviting, he bellows, "crunch bird, my ass!"

All of the action takes place in just 1 minute and 58 seconds, so when *The Crunch Bird* won the Oscar for Best Animated Short that year, it became the shortest film to ever receive an Academy Award in any category, a record it still holds to this day.

The Crunch Bird was significantly shorter than its two competitors that year. Over 10 delightful minutes, *Evolution* demonstrates how a single cell can evolve into a diverse array of lifeforms. At a runtime of 26 minutes, *The Selfish Giant* movingly tells the story of a giant who learns to be kind to children and is eventually rewarded with an eternity in paradise.

But Oscar voters were willing to overlook the time gap – and the difference in effort that comes with making a film 5 or 13 times as long. Is this normal for a film so short to win against longer animated films, or did *The Crunch Bird* buck the trend? Is the answer the same for live-action and documentary shorts? What about for feature-length films in the Best Picture race?

Since animated shorts tend to be a lot shorter than live-action ones, which in turn are briefer than documentary shorts, let's break this up by category:

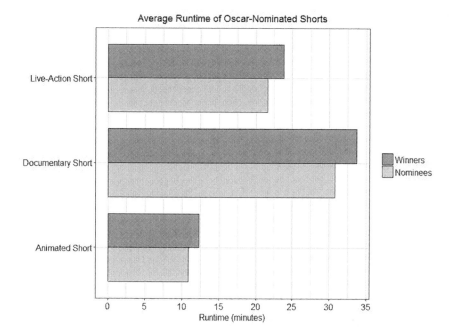

In all three short film categories (using data since 1975, the most recent year that runtimes are available for all nominees), the average winner runs between 1 and 3 minutes longer than the average nominee. While voters consider plenty of factors when deciding which movies to support, whether short or feature-length, the data suggests that there is a very slight correlation between Oscar short film success and runtime.

There are plenty of exceptions, of course. Since 1975, 12 animated shorts, 8 live-action shorts, and 4 documentary shorts have won the Oscar despite being the shortest nominees among their competitors. But in the other direction, 16 animated shorts, 13 live-action shorts, and 19 documentary shorts that were the longest entries of their season won the Oscar, so that seems to be a safer path to the podium.

It is a bit ironic that categories designed to honor short films still lean towards longer reels. But perhaps the true goal of a short film is to pack in as much emotion and storytelling as possible into a constrained amount of time, and it's easier to pack more in if you give yourself more time to play with.

There is a hard limit, though: Go over 40 minutes, and the Oscars classify a movie as a feature-length film, eligible for Best Picture but no longer in the race for the three short film categories. For movies over 40 minutes, do the results still hold? Do longer movies win Best Picture more often?

To answer this question, let's look at the runtime of every Best Picture winner and nominee in history, along with curves that show the trend over time:

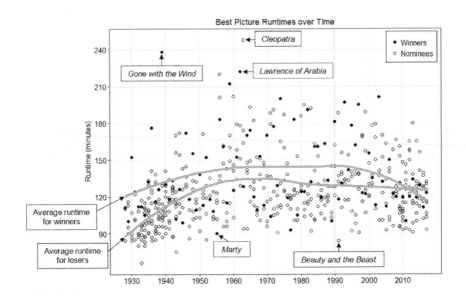

In the early years, Best Picture nominees were significantly shorter – both the winners and losers. Aside from *All Quiet on the Western Front* (1930), with its 2:32 runtime, no film in the first eight years of the Oscars reached the 2.5-hour mark. *The Great Ziegfeld* (1936) set the new record of 2:56. Three years later, *Gone with the Wind*

(1939) smashed that mark at 3:58 including musical interludes, still the record-holder for longest Best Picture champion.

After *Gone with the Wind*, it would be another 17 years before any nominees had the guts to break the 3-hour mark, when suddenly, three films passed that threshold all at once. In 1956, the year of the epic, Best Picture winner *Around the World in 80 Days* at 3:03, *Giant* at 3:21, and *The Ten Commandments* at 3:40 all went well over standard runtimes. The average across all five nominees that year was 2:55 per movie, a record which still stands today.

Things have calmed down a bit since then. Following the win by *The Lord of the Rings: The Return of the King* (2003), only *The Wolf of Wall Street* (2013) has reached the three-hour mark.

The graph shows that winners were generally longer than losers … until recently. Before the present decade, the average winner was 2:20, while the average of all other nominees was 2:02. In the current decade (that is, 2011-2017, up until this book went to press), the average winner runs for 1:59, while the average non-winning film clocks in at 2:07. It's too early to say whether this represents the start of a new trend towards shorter Best Picture winners.

Overall, the longest nominee has won 33 times, more than one out of every three years, while the shortest nominee has won only nine times, and just once since 1990: *Slumdog Millionaire* (2008) at an even 2:00.

Sometimes, the Academy loves a movie enough that it's willing to overlook its brevity. *Marty* (1955) won the award at just 90 minutes long, the shortest winner ever. *She Done Him Wrong* (1933) scored a nomination after just 66 minutes, still the shortest invitee. *Beauty and the Beast* (1991) made it onto the shortlist at 84 minutes, the shortest nominee since 1950.

The math says that for both short films and features, longer movies have won more frequently, but there may be a limit. *Cleopatra* (1963) went on for 4:08, the all-time record for a Best Picture nominee – and it lost to *Tom Jones* (1963), a movie half as long. And even *Tom Jones* was 65 times longer than *The Crunch Bird*, and that grisly little film managed to win an Oscar as well.

Chapter 14. Best Animated Feature

Do the Oscars have a Pixar bias?

> VALIANT:
>
> You crazy rabbit. I've been out there risking my neck for you. What are you doing? Singing and dancing!
>
> ROGER RABBIT:
>
> But I'm a toon. Toons are supposed to make people laugh.
>
> VALIANT:
>
> Sit down!
>
> ROGER RABBIT:
>
> You don't understand. Those people needed to laugh.
>
> –*Who Framed Roger Rabbit* (1988): Nominated for 6 Oscars; won 3 (Best Film Editing, Best Visual Effects, Best Sound Editing)

WHAT DO *Snow White and the Seven Dwarfs* (1937), *Fantasia* (1940), *Pinocchio* (1940), *101 Dalmatians* (1961), *Yellow Submarine* (1968), *Who Framed Roger Rabbit* (1988),[1] *Only Yesterday* (1991), *Ghost in the Shell* (1995), *Toy Story* (1995), *The Iron Giant* (1999), *Toy Story 2* (1999), and *Chicken Run* (2000) have in common?

These are the dozen animated films through the year 2000 that achieved higher than 95% on Rotten Tomatoes, a website that aggregates critic and audience reviews into scores that range from 0% to 100%.

You know what else they have in common? Not a single one was nominated for Best Picture. Nor was *The Jungle Book* (1967). Nor *The Little Mermaid* (1989). Nor *Aladdin* (1992). Nor *The Lion King* (1994). Nor dozens of other worthy contenders, perhaps overlooked because they employed animation instead of live action.

Only Disney's musical tale of an unlikely castle romance, *Beauty and the Beast* (1991), bucked the trend with an invitation to the Oscars' grandest category. In this century, *Up* (2009) and *Toy Story 3* (2010) also grabbed nominations, aided by the Academy doubling the size of the top category from five to ten nominees.

By the turn of the century, the Academy decided it was high time to honor animated achievement. With the genre finding no success in the Best Picture race, the Academy could at least create a separate category to honor animated features. This new category, introduced in 2001, coincided nicely with a flourishing of animated creativity from Hollywood and elsewhere. No longer just the domain of Disney, Best Animated Feature quickly developed into a hotly contested competition among a number of studios.

Let's break down the number of nominees and winners for Best

---

[1] *Who Framed Roger Rabbit* combines live action and animation to tell the story of a human detective and a cartoon bunny teaming up to save Los Angeles.

Animated Feature by production company to see how those studios have fared. Note that I assigned each film to whichever studio it is most commonly associated with, even when multiple studios played some role in releasing the film. A small number of films are assigned two studios.

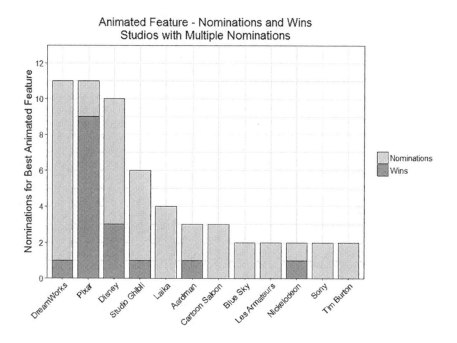

The immediate fact that jumps out of this graph is that three studios have dominated the Academy's newest category: DreamWorks and Pixar have 11 nominations apiece, and Disney has 10 nominations. No one else has more than 6.

But just as noteworthy are the rates at which these studios win: Pixar is a whopping 9-for-11, meaning that it has won in more than half of the 17 years that the category has existed. Disney and DreamWorks, despite earning a similar number of nominations as Pixar, only have 4 wins *combined*. So is Pixar making a far larger number of superior animated movies? Or do Academy voters have a bias in favor of Pixar?

To answer this question, we're going to need some measure of what the "right" call is. Admittedly, measuring film quality numerically is a fool's errand: My favorite film might be your least favorite, or vice versa, and neither of us is more correct than the other.

But what can be measured is how critics and audiences rate a movie. Using data from Rotten Tomatoes, we can look at average reviewer and user scores. These too are not perfect barometers of film quality. With respect to critics, Rotten Tomatoes' conversion of lengthy movie reviews into numeric scores is an imperfect science. With respect to users, the group of people who rate films online may not be representative of the moviegoing public at large.

Nevertheless, it's still a reasonable proposition to say that higher Rotten Tomatoes scores should be correlated with some sense of film quality.

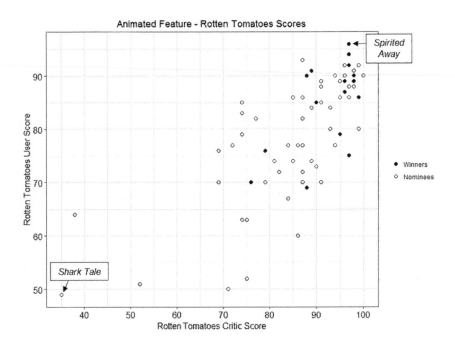

The vast majority of these points are in the upper-right corner of the graph, meaning that both critics and audiences liked the movie. In general, the Academy nominates animated features that both reviewers and the wider public enjoy. Even *Shark Tale*, which doesn't receive such lofty reviews, is still only one point shy of 50% according to users — maybe not quite Oscar-worthy, but ahead of plenty of non-nominated films with lower Rotten Tomatoes scores.

There is no right answer on whether critics or users are more correct when it comes to assessing film quality. That's a subjective question that each person can answer differently. So, I will weight them both equally to give each movie its overall Rotten Tomatoes score.

We can then judge all 17 years of Best Animated Features against these Rotten Tomatoes composite scores to determine what "should" have won. The results fall into these four categories:

1. The Rotten Tomatoes favorite won Best Animated Feature (8 times)

The following movies were all correct winners according to the Rotten Tomatoes composite score: *Spirited Away* (2001), *Finding Nemo* (2003), *The Incredibles* (2004), *WALL-E* (2008), *Up* (2009), *Inside Out* (2015), *Zootopia* (2016), and *Coco* (2017). Of those, six belong to Pixar, plus one win apiece for Studio Ghibli (*Spirited Away*) and Disney (*Zootopia*).

In these cases, when the winner matches the Rotten Tomatoes choice, we don't learn anything about a possible bias, because the Academy chose the presumed best film that year and wasn't necessarily biased by anything other than the quality of the movies.

2. No Pixar movies were nominated (4 times)

Years without any Pixar nominees also tell us nothing about a possible pro-Pixar bias or lack thereof. Aside from years we already addressed, there are four additional times when Pixar went uninvited to the Oscars:

- In 2005, Aardman's *Wallace and Gromit* not only saved vegetables from rabbits but also became the only stop-motion film to ever win Best Animated Feature. That year's highest-rated nominee was Hayao Miyazaki's antiwar film *Howl's Moving Castle*, produced by Studio Ghibli.

- In 2011, Nickelodeon's chameleon sheriff *Rango* won an Oscar, while Magic Light's *Chico & Rita* fell in love in pre-revolution Havana and came out ahead on Rotten Tomatoes.

- In 2013, Disney's *Frozen* built a snowman and a successful Oscar campaign, despite better reviews for Les Armateurs' bear-and-mouse duo *Ernest & Célestine*.

- In 2014, Disney won back-to-back trophies thanks to the robotics of *Big Hero 6*, though Cartoon Saloon's hand-drawn Celtic tale *Song of the Sea* outperformed it online.

3. Pixar should have won but lost (2 times)

- In 2001, the inaugural Best Animated Feature award went to DreamWorks' grudgingly heroic ogre *Shrek*, but the Rotten Tomatoes ratings preferred Pixar's loveable yet scary scream-gatherers at *Monsters, Inc.*

- A similar story unfolded in 2006: Animal Logic's penguins took home the prize for *Happy Feet*, but reviewers opted for Pixar's *Cars* driving through an old Route 66 town.

4. Pixar should have lost but won (3 times)

- It was a two-against-one contest in 2007, as Sony released the Iranian Revolution autobiography *Persepolis*, the best-reviewed animated film of the year according to Rotten Tomatoes, and also produced the animated mockumentary *Surf's Up*. But the lone Pixar nominee, *Ratatouille*, claimed the prize, the first of four in a row for Pixar.

- The last of those four straight, *Toy Story 3* (2010), theoretically should have lost to DreamWorks' *How to Train Your Dragon* (2010), if Rotten Tomatoes is to be taken literally. *Toy Story 3* earned a 98% from critics and an 89% from the general population for an average of 93.5%. *How to Train Your Dragon* scored 98% and 91% for an average of 94.5%. That puts *Dragon* ever so slightly ahead.

- In 2012, Disney's *Wreck-It Ralph* smashed Rotten Tomatoes to receive the highest average review score that year but was superseded at the Oscars by Pixar's Scottish adventure *Brave*.

Let's review the final standings. In 12 out of 17 years, the Oscars either made the right choice according to Rotten Tomatoes or had no Pixar nominees to choose from, so in those years there is no demonstration of either bias or no bias. In the remaining five years, the

Oscars awarded Pixar three times when the Rotten Tomatoes measure suggested a different result, and voted against Pixar twice when Rotten Tomatoes favored Pixar.

Add it all up and that gives us a 3-2 record in favor of Pixar's Oscar success relative to Rotten Tomatoes, certainly not a large enough gap to suggest that the Academy has a bias towards Pixar. Rather, Oscar voters seem to be frequently honoring a studio that's regularly producing high-quality cartoons.

Of course, this conclusion only holds if we accept Rotten Tomatoes as the true measure of underlying quality. In the next chapter, we'll look more broadly at the relationship between aggregated online scores and the Academy's verdicts.

Chapter 15. Best Documentary

Do the Oscars agree with critics and audiences?

> ADDISON:
>
> But more of Eve later. All about Eve, in fact. To those of you who do not read, attend the theater, listen to unsponsored radio programs, or know anything of the world in which you live, it is perhaps necessary to introduce myself. My name is Addison DeWitt. My native habitat is the theater. In it I toil not, neither do I spin. I am a critic and commentator. I am essential to the theater.

–All About Eve (1950): Nominated for 14 Oscars; won 6 (Best Picture, Best Director, Best Supporting Actor, Best Adapted Screenplay, Best Costume Design: Black-and-White, Best Sound)

THE MOST critical Best Picture decision the Academy made in the twenty-first century did not relate to any of the winners. It didn't even concern a nominee. It was in fact a decision *not* to nominate a certain movie: *The Dark Knight* (2008).

Experts and casual fans alike adored Christopher Nolan's epic *Batman* adaptation: Both groups give it an identical 94% rating – an excellent score – on Rotten Tomatoes. But professional and lay sentiments were not enough for the Academy, which instead nominated five films whose combined box office take was $100 million less than *The Dark Knight*'s billion-dollar haul.

Moviegoers shouldn't have been surprised by the omission. After all, *The Dark Knight* didn't receive Best Picture nominations from either the BAFTAs or the Golden Globes. Nevertheless, enough fans were outraged that the Academy decided to double the size of the category the following year in an effort to include more fan-friendly nominees.

But when it comes to picking winners, does the Academy make choices that critics or audiences approve of? To answer this question, I compiled four common measures of opinion – two for critics and two for audiences:

- **Metacritic:** This site converts critic reviews into ratings on a scale from 0-100, and then takes a weighted average of those ratings. The term "weighted average" means assigning more weight to some reviewers than others. Metacritic doesn't share these weights for every critic but says that they are "based on [the critics'] quality and overall stature."

- **Rotten Tomatoes critic score**: This metric is similar to Metacritic, except that Rotten Tomatoes simply marks each review as "fresh" or "rotten" – that is, positive or negative. Unlike Metacritic, Rotten Tomatoes assigns each review the same weight.

- **Rotten Tomatoes user score**: Rotten Tomatoes lets anyone on the internet rate a movie as "fresh" or "rotten," and displays the percentage of users who mark the movie as fresh.

- **Internet Movie Database (IMDb)**: This rating is similar to the Rotten Tomatoes user score in that anyone who goes to the website can rate a movie from 0-10 stars. IMDb adjusts the votes a bit – it doesn't explain the exact method – in order to prevent ballot stuffing.

I collected these four scores where available for every nominee in the categories of Best Documentary – the subject of this chapter – and Best Picture – which I include in as many chapters as possible due to its outsized importance.

As you might imagine, these four sets of scores are quite related to each other. To show this precisely, we can use a common statistical metric called correlation, which is similar to the colloquial definition of the word "correlation." For this metric, a 0 means that no relationship exists between two sets of data, a 1 means that the two sets agree perfectly, and the higher the number, the more agreement between the two sets.

	Metacritic	Rotten Tomatoes Critics	Rotten Tomatoes Users	IMDb
Metacritic		.65	.26	.32
Rotten Tomatoes Critics	.65		.46	.43
Rotten Tomatoes Users	.26	.46		.81
IMDb	.32	.43	.81	

The most related pair are Rotten Tomatoes user scores and IMDb user scores at 0.81 correlation. The second-most related pair are Rotten Tomatoes critic scores and Metacritic scores at 0.65 correlation. This is not surprising, given that Rotten Tomatoes user scores and IMDb are measuring the same thing – audience opinion – and Rotten Tomatoes critic scores and Metacritic are likewise measuring the same thing – critical opinion.

To answer our original question about how well the Oscars correlate with reviewer and fan opinions, we only require one measure each for audience and critic opinions, because we don't need two numbers telling us the same fact (how much reviewers or fans liked a certain movie). So, we'll go with the ones that cover more movies. For users, IMDb rates more Best Picture and Best Documentary nominees than Rotten Tomatoes. For critics, Rotten Tomatoes covers more of these films than Metacritic.

Critics and audiences generally agree in their film tastes, but there are numerous exceptions. In the graph below showing all Best Picture and Best Documentary nominees since the Oscars began, movies farther to the right are the most lauded by critics. Movies higher on the chart received the highest praise from audiences. We can see that the farther to the right a movie is, the more likely it's also higher up, but it's certainly not a straight line indicating perfect agreement.

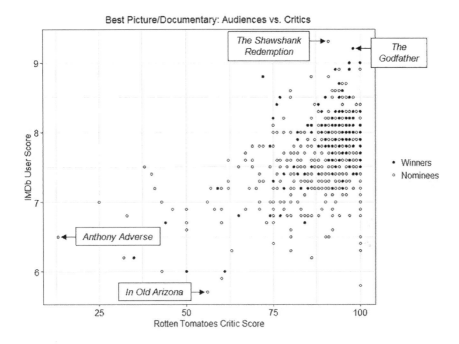

The very highest audience rating, 9.3, belongs to *The Shawshank Redemption* (1994), which took part in the greatest Best Picture year of all time according to IMDb users, with an average rating of 8.3 across the five contenders. The other nominees that year were *Four Weddings and a Funeral*, *Pulp Fiction*, *Quiz Show*, and the movie that beat all of them at the Oscars, *Forrest Gump*. The highest audience-rated Best Picture winner on IMDb – since *Shawshank* lost – is *The Godfather* (1972) at 9.2, followed by *The Godfather: Part II* (1974) at 9.0.

For documentary nominees, the honor of highest IMDb rating goes to both *Front Line* (1979) and *O.J.: Made in America* (2016) at 9.0. *Front Line* – about a photographer in the Vietnam War – only has 19 reviews, while *O.J.: Made in America* – ESPN's Oscar-winning miniseries about O.J. Simpson – has over 14,000 ratings.

Among Rotten Tomatoes critics, five Best Picture winners enjoy a perfect score: *All Quiet on the Western Front* (1930), *Rebecca* (1940), *The Lost Weekend* (1945), *All About Eve* (1950), and *Marty* (1955). These five champions are joined by seven documentary winners which also earned a 100% on Rotten Tomatoes: *Woodstock* (1970), *Marjoe* (1972), *Hôtel Terminus* (1988), *American Dream* (1990), *Taxi to the Dark Side* (2007), *Man on Wire* (2008), and *O.J.: Made in America* (2016).

On the other end of the spectrum, IMDb users consider *In Old Arizona* (1928) at 5.7 to be the worst Best Picture nominee ever, while *Cimarron* (1931) and *Cavalcade* (1933) are tied for the worst winner ever at 6.0. Rotten Tomatoes critics pick *Anthony Adverse* (1936) at 13% as the worst Best Picture nominee of all time and *The Broadway Melody* (1929) at 35% as the worst winner. Granted, earlier movies tend to have fewer reviews on these sites, which can lead to more extreme scores, both good and bad.

According to both users and critics, the year of *The Broadway Melody*, 1929, was the worst in Best Picture history. However, the data is incomplete for 1929. One of that year's competitors, *The Patriot*, is the only Best Picture nominee that has since been lost and therefore can't be graded by modern websites. After 1929, users say the second-worst year was 1931 (*Cimarron* won; *East Lynne*, *The Front Page*, *Skippy*, and *Trader Horn* were nominated), while critics hand that dubious honor to 1969 (*Midnight Cowboy* won; *Anne of the Thousand Days*, *Butch Cassidy and the Sundance Kid*, *Hello, Dolly!*, and *Z* were nominated).

To see whether winners have higher IMDb and Rotten Tomatoes scores than losers, let's compare the average scores of winners and losers in both Best Documentary and Best Picture, only using nominees for years in which every nominee is reviewed by the website in question.

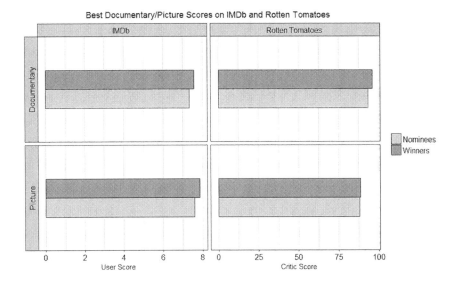

In all four boxes, the dark grey bar (the average IMDb user score or Rotten Tomatoes critic score for winners) is just a bit longer than the light grey bar (the average score for nominees that lost the Oscar). The differences are tiny – between 0.2 and 0.3 on IMDb's scale, and between 0.5% and 3.0% on Rotten Tomatoes' scale. But they do suggest that the Oscars slightly prefer movies that critics and audiences appreciate.

Despite this modest trend, the Oscars disagree with both critics and audiences every so often. In fact, there have been eight Best Picture winners that both IMDb and Rotten Tomatoes consider to be the single worst nominee of the year: *Cimarron* (1931), *Cavalcade* (1933), *Gentleman's Agreement* (1947), *The Greatest Show on Earth* (1952), *Around the World in 80 Days* (1956), *Gigi* (1958), *Oliver!* (1968), and *Chariots of Fire* (1981).

Fans of shows like the People's Choice Awards – which let audiences vote on the year's best in cinema – may find it a little odd to let a group of approximately 8,000 people primarily based in Los

Angeles choose the best movie of the year, when millions of moviegoers around the world don't get a say. But they should be reassured to know that the opinions of those 8,000 industry insiders are correlated with public opinion. Sometimes, when those Academy voters disagree with the public, the Academy is eventually overruled by the judgment of history, as we'll see in Chapter 24. Other times, the voters disagree with the audience but are praised in hindsight for sticking to their convictions.

Chapter 16. Best Foreign Language Film

Do film festivals track the Oscars?

> FOGG:
>
> My usual routine is beside the point. We're leaving for the Continent in ten minutes.
>
> PASSEPARTOUT:
>
> Monsieur is going traveling?
>
> FOGG:
>
> Yes. Around the world.
>
> PASSEPARTOUT:
>
> Then you will not be here for breakfast.

–*Around the World in 80 Days* (1956): Nominated for 8 Oscars; won 5 (Best Picture, Best Adapted Screenplay, Best Original Score, Best Cinematography: Color, Best Film Editing)

Film festivals have a rather sordid origin story. The first-ever film festival began innocently enough in 1932 as a noncompetitive event sponsored by the arts organization Venice Biennale. Almost immediately, however, Italy's fascist government transformed the young event into propaganda for the regime and its dark ideology.

By the second Venice Film Festival in 1934, new awards for Best Italian Film and Best Foreign Film were christened the "Mussolini Cup" after dictator Benito Mussolini. The awards for Best Actor and Best Actress were dubbed "Great Gold Medals of the National Fascist Association for Entertainment."

Over the years, the soon-to-be Allied nations grew exasperated that a soon-to-be Axis country was in charge of the world's preeminent movie gathering. In response, the French introduced a film festival in Cannes in 1939, which was to be headlined by a new American film called *The Wizard of Oz*. Or at least that was the plan. The opening night of the inaugural festival was scheduled for September 1, which happened to be the very day that Hitler invaded Poland, sparking World War II. The festival was cancelled. Two days later, France and the United Kingdom declared war on Germany.

After the world's deadliest nightmare concluded in 1945, France revived the idea of a Cannes Film Festival. On September 20, 1946, the Free World finally hosted its first film festival. It was not exactly smooth sailing: One film was projected upside down; another was shown in reverse order. Nevertheless, the crowd was awed by such illustrious films as *Gaslight*, the thriller that coined the term for tricking someone into doubting his or her own sanity; *Rome, Open City*, a glimpse of Rome under Nazi occupation; and the Oscars' Best Picture winner *The Lost Weekend*, a battle between a man and his alcoholism. The festival was a hit, and a new tradition was born.

West Berlin added its own film festival five years later, rounding out the trifecta that is commonly known as the Big Three. Since then, hundreds of other cities have launched film festivals of their own, and the festival circuit has blossomed into a prominent publicity opportunity for both movies and the people who make them.

One of the chief reasons that filmmakers crave festival recognition is to get a head start in the Oscar jockeying. In particular, for foreign films that don't receive a lot of screen time in Los Angeles, festivals are considered a smart pathway to attracting Oscar buzz.

While math can't answer whether or not film festival honors *cause* movies to win Oscars, we can use historical data to determine whether these two events are *correlated* – that is, whether the two groups like the same films more often than random chance would suggest. Since there are hundreds of film festivals whose lineups have almost no overlaps with the list of Oscar nominees in any given year, we will simplify the analysis by restricting ourselves to the festivals accredited by the International Federation of Film Producers Associations, the chief regulating body for international film festivals. Ignoring specialized film festivals, such as those that only consider movies of one specific genre, we are left with 17 of the world's most renowned festivals:

Not all of these 17 festivals are created equal in terms of correlation with the Oscars. I gathered data on how often the winner of each festival's most prestigious prize goes on to be nominated for or win the Oscar for Best Foreign Film. Using data from 1956 (the first year that the Oscars presented a competitive Foreign Language Film award) to the present, here are the number of overlaps between film festivals and the Academy Awards:

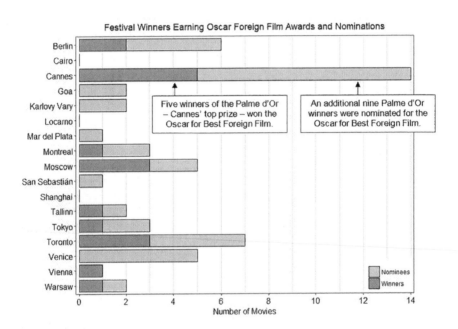

By far the closest Oscar parallel among these festivals is Cannes, hardly a surprise given its lead in name recognition. Toronto, Moscow, and Berlin form the next tier of Oscar correlation, while Venice has honored five Oscar nominees but never an Oscar champion.

On the other end of the spectrum, three of these festivals – Cairo, Locarno, and Shanghai – have never awarded a single Oscar nominee.

In order to determine whether film festival winners claim Oscars more often than non-festival winners, let's divide Oscar years into groups based on how many of that year's Best Foreign Film Oscar nominees won a film festival. The years with zero festival champions competing at the Oscars don't help us answer the question, so we'll throw those years out. First up – the 23 years when exactly one film festival winner was nominated for Best Foreign Film:

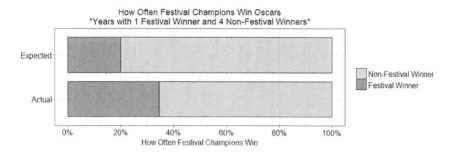

The first bar, labeled "Expected," shows how often we would expect film festival winners to also win Best Foreign Film at the Oscars if festival status didn't have an impact. Since this chart covers years in which one out of five Oscar nominees was a festival winner, the expected bar shows a 20% winning rate (that is, one-fifth) for festival winners.

The second bar, labeled "Actual," shows what actually happened in these 23 years with exactly one festival winner competing in Hollywood. In practice, a festival winner emerged victorious at the Academy Awards in 35% of these years, which is higher than 20%. This piece of evidence suggests that festival winners may have an advantage over Oscar nominees that do not have a festival title on their résumés.

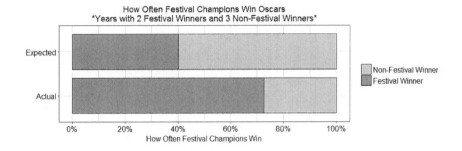

In the 11 years in which two festival champions went head-to-head at the Oscars, we would expect 40% (or two-fifths) of years to award an Oscar to a movie with a festival honor. But once again, the "Actual" bar is longer than the "Expected" bar for festival winners. In this case, 73% of years saw one of the two competing festival winners also take home the trophy in Los Angeles.

That 73% represents 8 of 11, meaning there are only three years when two festival winners went up against each other at the Oscars and both lost to a film that didn't win one of the major festivals included in this chapter:

- In 1986, Netherland's *The Assault* by Fons Rademakers beat the festival winners from both Montreal (France's *Betty Blue* by Jean-Jacques Beineix) and Toronto (Canada's *The Decline of the American Empire* by Denys Arcand).

- In 1993, Spain's *Belle Epoque* by Fernando Trueba beat the festival winners from both Berlin (Taiwan's *The Wedding Banquet* by Ang Lee) and Cannes (Hong Kong's *Farewell My Concubine* by Chen Kaige).

- In 2017, Chile's *A Fantastic Woman* by Sebastián Lelio beat the festival winners from both Berlin (Hungary's *On*

Body and Soul by Ildikó Enyedi) and Cannes (Sweden's *The Square* by Ruben Östlund).

We've now covered years with one festival winner at the Oscars and years with two festival winners at the Oscars. Only twice have three festival winners crowded the Academy Awards field:

- In 2000, Taiwan's *Crouching Tiger, Hidden Dragon* by Ang Lee won Toronto, Mexico's *Amores Perros* by Alejandro G. Iñárritu won Tokyo, and France's *The Taste of Others* by Agnès Jaoui won Montreal. *Crouching Tiger* won the Foreign Film Oscar (and was nominated for Best Picture).

- In 2009, Israel's *Ajami* by Scandar Copti and Yaron Shani won Tallinn, Peru's *The Milk of Sorrow* by Claudia Llosa won Berlin, and Germany's *The White Ribbon* by Michael Haneke won Cannes. All of them lost the Oscar to Argentina's *The Secret in Their Eyes* by Juan José Campanella.

Two years of data – 2000 and 2009 – isn't enough to draw conclusions from. So, our most useful data comes from the 23 years when one festival winner competed at the Oscars, and the 11 years when two festival champions were Oscar nominees. And from those 34 years, it does appear that movies taking top honors on foreign soil are more likely to repeat in America.

So all the glamour of far-off film festivals isn't just for fun – it's also for the serious business of trying to move ahead in the Oscar race.

Chapter 17. Best Popular Film

Which past movies would have won Best Popular Film?

CHARLEY:
Tell 'em that the New York State Supreme Court rules there's no Santa Claus. It's all over the papers. The kids read it and they don't hang up their stockings. Now, what happens to all the toys that are supposed to be in those stockings? Nobody buys 'em. The toy manufacturers are gonna like that. So they have to lay off a lot of their employees. Union employees. Now you got the C.I.O. and the A. F. of L. against you. And they're gonna adore you for it. And they're gonna say it with votes. Oh, and the department stores are gonna love you too, and the Christmas card makers and the candy companies. Oh Henry, you're gonna be an awful popular fella.

–*Miracle on 34th Street* (1947): Nominated for 4 Oscars; won 3 (Best Supporting Actor, Best Adapted Screenplay, Best Story)

August 8, 2018, appeared to be a seminal day in Oscar history. That morning, the Academy announced three major changes:

- Starting in 2020, the Oscar ceremony will be held on an earlier date.

- To shorten the ceremony to three hours, some categories will be announced during commercial breaks rather than on live television.

- The Academy is adding a category to honor achievement in popular film.

I had a lot of thoughts right off the bat, but one of them was, "I'm nearly finished writing a book on the Oscars, with one chapter per category – and for the first time since Best Animated Feature was introduced in 2001, the Academy just added a new category!"

I dashed off a few tweets and then drove to work, spending the L.A. car ride thinking about this new popular film category. What would the rules be? Which Academy members would decide the nominees? What constitutes a popular film? What is the best formula to predict this new category? How would previous Oscars have played out differently had there been a Best Blockbuster race?

It turns out that the Academy itself hadn't completely thought through these questions either. One month later, the Board of Governors announced that the Academy was putting the new popular film category on hold, at least for now.

Nevertheless, it remains a fascinating thought experiment, especially if the Academy returns to the idea of a Best Popular Film category in the future, as the Board has suggested it might. Which movies would have won Best Popular Film from 1928 to 2017, according to Oscar voters?

We don't need a *Back to the Future*-style time machine to rewrite history and answer this. We can surmise imaginary winners of Best Popular Film with a combination of educated guesswork, logic, and math.

Without knowing what rules the Academy would create, let's say that Best Popular Film is defined as the best movie of the year among the top 20 domestic box office earners. I will assume that the voters are honoring only quality, not revenue, so the best film is simply the best film: If the Best Picture winner is among those top 20, it would have also won Best Popular Film.

If the Best Picture winner did not crack that year's top 20 in ticket sales, my next assumption is that the Academy would have chosen a Best Picture nominee that did. If multiple Best Picture nominees made the cut, I will assume the one with the highest probability to win Best Picture according to my Oscar model was the Academy's favorite and would have won Best Popular Film. If not a single Best Picture nominee came in the top 20 at the box office, I'll resort to calculating which film would have had the best chance of winning Best Picture, had it been nominated.

One caveat: In the real world, only revenue earned prior to the nominations announcement would determine eligibility for this fictional category. That said, for many earlier years it is difficult to break down revenue by time period, so I am using total revenue for each nominee. Since the Oscars generally fall two or more months after the end of the calendar year, most of these films earned the vast majority of their revenue before the Oscars.

With these rules in place, here are the completely imaginary winners of Best Popular Film from 1928-2017 (Best Picture winners appear in bold):

1928: *Wings*
1929: *The Broadway Melody*
1930: *All Quiet on the Western Front*
1931: *Cimarron*
1932: *Grand Hotel*
1933: *Cavalcade*
1934: *It Happened One Night*
1935: *Mutiny on the Bounty*
1936: *The Great Ziegfeld*
1937: *The Life of Emile Zola*
1938: *You Can't Take It with You*
1939: *Gone with the Wind*
1940: *Rebecca*
1941: *How Green Was My Valley*
1942: *Mrs. Miniver*
1943: *Casablanca*
1944: *Going My Way*
1945: *The Lost Weekend*
1946: *The Best Years of Our Lives*
1947: *Gentleman's Agreement*
1948: *Hamlet*
1949: *All the King's Men*
1950: *All About Eve*
1951: *An American in Paris*
1952: *The Greatest Show on Earth*
1953: *From Here to Eternity*
1954: *On the Waterfront*
1955: *Picnic*
1956: *Around the World in 80 Days*
1957: *The Bridge on the River Kwai*

1958: *Gigi*
1959: *Ben-Hur*
1960: *The Apartment*
1961: *West Side Story*
1962: *Lawrence of Arabia*
1963: *Tom Jones*
1964: *My Fair Lady*
1965: *The Sound of Music*
1966: *A Man for All Seasons*
1967: *Guess Who's Coming to Dinner*
1968: *Oliver!*
1969: *Midnight Cowboy*
1970: *Patton*
1971: *The French Connection*
1972: *The Godfather*
1973: *The Sting*
1974: *The Godfather: Part II*
1975: *One Flew over the Cuckoo's Nest*
1976: *Rocky*
1977: *Annie Hall*
1978: *The Deer Hunter*
1979: *Kramer vs. Kramer*
1980: *Ordinary People*
1981: *Chariots of Fire*
1982: *Gandhi*
1983: *Terms of Endearment*
1984: *Amadeus*
1985: *Out of Africa*
1986: *Platoon*
1987: *Moonstruck*

1988: *Rain Man*
1989: *Driving Miss Daisy*
1990: *Dances with Wolves*
1991: *The Silence of the Lambs*
1992: *Unforgiven*
1993: *Schindler's List*
1994: *Forrest Gump*
1995: *Braveheart*
1996: *The English Patient*
1997: *Titanic*
1998: *Shakespeare in Love*
1999: *American Beauty*
2000: *Gladiator*
2001: *A Beautiful Mind*
2002: *Chicago*
2003: *The Lord of the Rings: The Return of the King*
2004: *Million Dollar Baby*
2005: Walk the Line
2006: *The Departed*
2007: Juno
2008: *Slumdog Millionaire*
2009: Avatar
2010: *The King's Speech*
2011: The Help
2012: Lincoln
2013: Gravity
2014: American Sniper
2015: The Revenant
2016: La La Land
2017: Dunkirk

Right off the bat, it's easy to spot why the Academy did not introduce a Best Popular Film category for nine decades, but now believes one may be warranted. From 1928 to 2004, the Best Picture winner would have also won Best Popular Film by these rules in all but two years:

- In 1955, the charming story of *Marty*, a shy butcher falling in love with an equally shy schoolteacher, captured the hearts of the Academy, becoming the first film ever to fall outside the year's top 20 at the box office but still win Best Picture. Among Best Picture nominees that year, only *Mister Roberts* and *Picnic* would have been eligible for Best Popular Film. While *Mister Roberts* did better at the box office, it's pretty clear which movie the Academy preferred: *Picnic* was nominated by the Directors Guild and the Oscars for Best Director, and it won Best Film Editing. *Mister Roberts* wasn't nominated for any of those and had three fewer total nominations. So in all likelihood, Oscar voters would have chosen *Picnic* for Best Popular Film, awarding the steamy Kansas romance an additional trophy.

- In 1987, the cutoff for top 20 status was $50 million, and Best Picture winner *The Last Emperor* just missed the mark at $44 million. That would have left Best Popular Film up for grabs, with Best Picture nominees *Broadcast News*, *Fatal Attraction*, and *Moonstruck* all eligible. This one would have been very close. *Broadcast News* had the most nominations (seven, compared to six for the other two films). *Fatal Attraction* was the most commercially successful of the three. But on Oscar evening, those two films were shut out completely. The math says that

Moonstruck, which won Oscars for Best Original Screenplay, Best Actress (Cher), and Best Supporting Actress (Olympia Dukakis), would have won its fourth Oscar in the Best Popular Film category.

Then, in 2005, everything changed. Just two years after the extremely popular *The Lord of the Rings: The Return of the King* (2003) swept the Oscars, the Academy did not nominate a single top 20 box office film for Best Picture – the only time in history that has occurred. The model chose *Walk the Line* (2005) as the most likely Best Picture winner among the year's top earners, had it been nominated, thanks to its Golden Globe victory, Best Actress win (Reese Witherspoon), Best Actor nomination (Joaquin Phoenix), and Best Film Editing nomination.

From that point on, 10 of the 13 most recent years have a Best Picture winner that, by this chapter's standards, would not have been eligible for Best Popular Film. Even the three Best Picture winners in this era that would have also won Best Popular Film – *The Departed* (2006), *Slumdog Millionaire* (2008), and *The King's Speech* (2010) – weren't exactly blockbusters.

Juno (2007), *The Help* (2011), and *American Sniper* (2014) were the only Best Picture nominees of their respective years eligible for Best Popular Film by my criteria, so I marked them as the hypothetical Best Popular Film winners by default. Presumably, the Academy would vote for Best Picture nominees ahead of non-Best Picture nominees.

But all of this comes with the qualifier "hypothetical." If the Academy revives this idea, it could easily establish different rules for determining what constitutes a popular film. The voters could include popularity, and not just quality, in their considerations, perhaps using box office numbers. That might mean a Best Picture winner like *Slumdog Millionaire* could lose Popular Film to a non-Best Picture nominee like *The Dark Knight* even if *Slumdog Millionaire* were eligible in both categories. Additionally, the Academy could simply declare that Best Picture winners will be ineligible for Best Popular Film.

But if the Board of Governors is hunting for cold, hard evidence that Oscar winners are growing less popular, they need look no further than this chapter. We saw three quarters of a century of agreement between audiences and Oscar voters, followed by a dramatic turnaround in 2005.

What this data does not tell us is which side diverged from the other. Audiences might accuse the members of the Academy of becoming too pretentious in their tastes. Oscar voters might retort that the average moviegoer has turned to tired sequels and popcorn flicks over true art. Maybe both of those statements have an element of truth mixed with hyperbole.

Will the 2020s see the introduction of Best Popular Film, and if so, will that help to lessen the gap, or only serve to increase the divide? Stay tuned.

Chapter 18. Best Original Screenplay

How important is Writers Guild Award eligibility?

>   GILLIS:
> She was so like all us writers when
> we first hit Hollywood - itching with
> ambition, panting to get your names up
> there: Screenplay by. Original Story
> by. Audiences don't know somebody sits
> down and writes a picture. They think
> the actors make it up as they go along.
>
> –*Sunset Boulevard* (1950): Nominated for 11 Oscars; won 3 (Best Original Screenplay, Best Original Score, Best Art Direction: Black-and-White)

GOLD DERBY, a website that covers awards shows, has run the same headline every year from 2013 to 2018: "WGA is limited as guidepost to predicting Oscars." It occasionally substitutes in "Writers Guild Awards" for "WGA" (which officially stands for Writers Guild of America) or rearranges the title, but the words "limited," "guidepost," and "Oscars" always appear.

The article invariably begins by bemoaning the fact that the Writers Guild annually deems some films ineligible for WGA recognition. Specifically, the Writers Guild often does not consider animated pictures, foreign films, or scripts written by non-guild members when choosing its five nominees for Best Original Screenplay and five for Best Adapted Screenplay. Consequently, seven Oscar screenplay winners since 1984, the year that the Writers Guild switched to its current category format, never had a chance to compete at the WGA. From a forecasting perspective, I understand Gold Derby's frustration, but at the same time, the Writers Guild's primary purpose is to further the interests of its membership.

Despite the WGA's peculiarities, its awards have proven too valuable to ignore for Oscar prediction purposes. Between 1984 and 2017, the Writers Guild Awards and the Academy Awards have each honored 68 screenplays apiece: 34 original scripts in one category and 34 scripts adapted from previously existing works in another. Of those 68, they agreed on 45, or 66%. In the oft-inaccurate world of Oscar precursors, 66% is nothing to scoff at. By way of comparison, the BAFTAs are only at 37% accuracy in these categories over that same time frame.

So, we can't just throw out the WGA results. We have to figure out how to handle the problem of WGA eligibility in Oscar prediction. To start, we can split all Oscar screenplay nominees into four groups:

- Movies that won a Writers Guild Award

- Movies that were nominated by the WGA but lost

- Movies that were eligible for a WGA nomination but did not receive one

- Movies that were ineligible for a WGA nomination

At first blush, it might be tempting to lump the last two groups together. After all, none of the films in either group were nominated for or won a Writer Guild Award. The trouble is, they mean very different things. If a script is eligible for a WGA nomination but does not receive one, that omission sends a clear message: The Writers Guild voters did not consider the movie one of the top five original or top five adapted screenplays of the year. That's a bad omen for the film's Oscar chances. But if the script is ineligible, then its lack of a WGA nomination doesn't tell us anything – good or bad – about a movie's chances.

Obviously, it's best for a script's Oscar chances to win a Writers Guild Award. But what about that second category, films that are nominated but lose? On the one hand, a nomination is indisputably a compliment. On the other hand, the Writers Guild has declared the movie inferior to at least one other nominee.

Historically, it turns out that nominees who don't win the WGA perform worse at the Oscars than Oscar nominees who were ineligible for the WGA:[1]

---

[1] The first pie chart says that 68% of WGA winners also win the Oscar, but the text earlier stated that figure as 66%. The reason for this is that the chart only includes movies that were nominated for the Oscar. *Roxanne* (1987) and *Bowling for Columbine* (2002) won Writers Guild Awards but did not receive Oscar screenplay nominations.

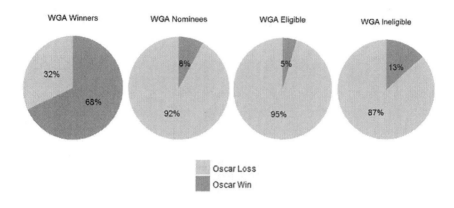

Unsurprisingly, the worst of all scenarios for an Oscar hopeful is to be eligible for the Writers Guild Award but not nominated for it. Only two movies have ever overcome this predicament to win a screenplay Oscar: *Amadeus* (1984) and *The Usual Suspects* (1995).

When I predict the Oscars, I use historical data – the same data that goes into those pie charts – to determine how the WGA results affect the Oscars. My model considers how many nominees that year fall into each WGA group: winner, nominee, not nominated, and not eligible. A five-nominee category actually presents 46 possible scenarios, but let's focus on the ones that occur most frequently.

To start, there is typically one and only one WGA winner nominated in each Oscar screenplay category. The last time a screenplay won a Writers Guild honor but wasn't even nominated for a writing award at the Oscars was *Bowling for Columbine* (2002). And the only time that more than one WGA winner competed in one Oscar category was the 2016 Best Adapted Screenplay race, when both *Arrival* and *Moonlight* (the eventual winner) were up for the award. Tarell Alvin McCraney and Barry Jenkins based the screenplay of *Moonlight* on McCraney's unproduced play, which made the script "original" by the WGA's standards but "adapted" by the Academy's.

Furthermore, in 93% of races, the WGA and the Academy have at least three overlaps in their choices of nominees.

These filters described in the last two paragraphs whittle us down to the six most common scenarios, listed alongside what my model says are the chances that each film wins a screenplay Oscar:

Winners	Nominees	Eligible	Ineligible	Winner %	Nominee %	Eligible %	Ineligible %
1	2	0	2	64%	6%		12%
1	2	1	1	69%	7%	4%	13%
1	2	2	0	74%	8%	5%	
1	3	0	1	67%	7%		12%
1	3	1	0	72%	8%	4%	
1	4	0	0	70%	8%		

Here's how to read this chart. Each row is a common scenario, and the scenario is defined in the first four columns. Note that these first four columns always add up to five, because the Academy always nominates five screenplays per category.

For example, the first row represents the scenario in which one of the Oscar nominees won a Writers Guild award, two other Oscar nominees were nominated for a WGA award, zero Oscar nominees were WGA-eligible without being nominated by the WGA, and two Oscar nominees were WGA-ineligible.

Then, the last four columns show the probability that one of the movies of each WGA status wins the Oscar. In the first row, the WGA winner has a 64% chance to win Best Screenplay. Each of the two films that lost at the Writers Guild Awards has a 6% chance to win the Oscar. And each of the two WGA-ineligible films has a 12% chance to win.

On average, the WGA winner has a 69% chance to win the Oscar, but it varies based on the statuses of the opponents. In that first row, the WGA winner only has a 64% chance to win because it has to go up against two movies that it didn't have to compete against at the Writers Guild contest. But in the third row, the competition is considerably weaker: Two WGA-eligible Oscar nominees didn't even get nominated by the WGA. The WGA winner's Oscar chances climb all the way to 74%.

With this chart, the relative ranking of WGA statuses is evident: Winning a Writers Guild trophy is far and away the best-case scenario for an Oscar hopeful. Being ineligible for a Writers Guild award, while nowhere near as nice as winning one, is the next-best scenario because at least the film didn't lose. Being nominated but losing a Writers Guild nomination isn't ideal, but it's still better than not getting nominated, the worst-case situation.

The Writers Guild results don't entirely eliminate an ineligible or losing film's chances – just ask *Amadeus*. Or better yet, ask *The Usual Suspects* – there's a film that knows a thing or two about pulling off surprise endings.

Chapter 19. Best Adapted Screenplay

Which source materials win Oscars?

                POPESCU:

I'd say you were doing something pretty dangerous this time.

                MARTINS:

Yes?

                POPESCU:

Mixing fact and fiction.

                MARTINS:

Should I make it all fact?

                POPESCU:

Why no, Mr. Martins. I'd say stick to fiction. Straight fiction.

—*The Third Man* (1949): Nominated for 3 Oscars; won 1 (Best Cinematography: Black-and-White)

In 1961, screenwriter Abby Mann broke an Oscar curse that he himself may not have been aware existed. In winning Best Adapted Screenplay for *Judgment at Nuremberg*, a powerful window into American efforts to bring Nazi judges and lawyers to justice, Mann wrote the first nonfiction script in 24 years to win the category.

Previously, only *The Life of Emile Zola* (1937) had won Best Adapted Screenplay for a work of nonfiction.[1] Other winning scripts had elements of truth in them, such as *The Lost Weekend* (1945) and *The Bridge on the River Kwai* (1957), but both changed enough names and events to fall squarely within the realm of fiction.

When Mann accepted his Oscar, he said, "A writer who's worth his salt at all has an obligation not only to entertain but to comment on the world in which he lives." His words must have touched a nerve because suddenly the Academy couldn't get enough of nonfiction scripts. *Becket* (1964), *A Man for All Seasons* (1966), *The Lion in Winter* (1968), and *The French Connection* (1971) went on to triple the number of nonfiction Adapted Screenplay winners in Oscar history within the course of a decade.

But even within fiction and nonfiction, there is a great variety of inspiration from which to choose. Fictional movies can be based on anything from plays to short stories to, in one case, an epic poem: *O Brother, Where Art Thou?* (2000) is a take on Homer's *The Odyssey* set in Mississippi during the Great Depression. Nonfiction can derive from sources such as books, newspaper articles, and biographies.

---

[1] *The Story of Louis Pasteur* (1936) also won both story and screenplay honors for its biography of the famed French biologist. The categorization was different back then, so it was able to take home two Oscars including the forerunner to the modern Adapted Screenplay, but probably would have competed for Original Screenplay under today's rules.

Does the Academy have a preference? Is there an advantage to adapting a screenplay from a certain type of source material? Let's take a look at the data:

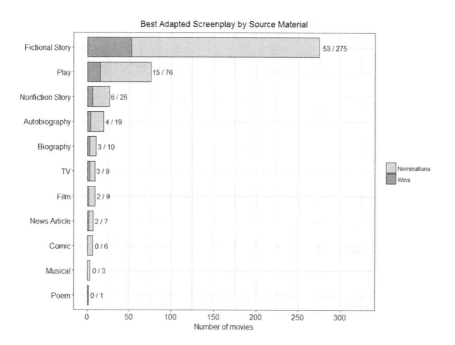

Fictional stories, such as novels, novellas, and short stories, are by far the most popular source material for both nominees and winners. This may reflect an industry bias towards writing films based on novels, or an Academy preference for nominating such films, or perhaps some of both.

But to answer our question on which source materials have better chances of winning, we care more about winning percentages than simple totals. The four categories in the chart with more than ten nominees all have remarkably similar winning percentages, falling between 20% and 23%. So, the Academy doesn't seem to have a strong preference among these types of sources.

In the smaller-sample categories, we can't draw any firm conclusions, but it's worth noting that poetry (*O Brother, Where Art Thou?*), musicals, and comics are collectively zero-for-ten, still looking for their first win. The three that derive from stage musicals are *Oliver!* (1968), *Cabaret* (1972), and *Chicago* (2002). The six based on comics, graphic novels, or related materials are *Skippy* (1931), *Ghost World* (2001), *Shrek* (2001), *American Splendor* (2003), *A History of Violence* (2005), and *Logan* (2017).

These averages, however, are far from constant over time. In order to visualize this change over the decades, let's group the narrow categories listed above into three broader categories:

- **Fiction:** Comic, Fictional Story, Poem
- **Nonfiction:** Autobiography, Biography, News Article, Nonfiction Story
- **Performance:** Film, Musical, Play, TV

We can see a dramatic increase in the number of nonfiction nominees per year in this century:

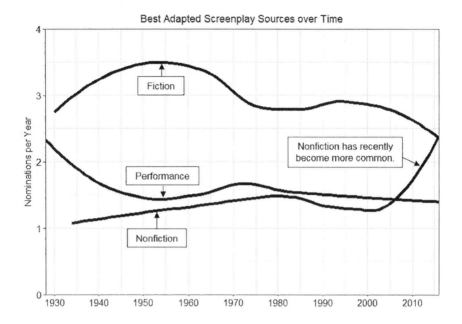

Of course, grouping all nominees by source material doesn't give proper credit to the differences in quality within a given source material. Yes, 76 plays have been converted into Oscar-nominated scripts, but some plays are simply better than others. As a result, it's no surprise that some authors have repeatedly seen their works adapted for the big screen. The record for the most works by one author adapted into Best Adapted Screenplay nominees is three:

Authors who sourced three Adapted Screenplay nominees (winners in bold):

- Sinclair Lewis: *Arrowsmith* (1931), *Dodsworth* (1936), ***Elmer Gantry*** (1960)

- Edna Ferber: ***Cimarron*** (**1931**), *Stage Door* (1937), *Giant* (1956)

- Dashiell Hammett: *The Thin Man* (1934), *After the Thin Man* (1936), *The Maltese Falcon* (1941)

- Lillian Hellman: *The Little Foxes* (1941), *Watch on the Rhine* (1943), ***Julia*** **(1977)**

- Charles Dickens: *Great Expectations* (1946), *Oliver!* (1968), *Little Dorrit* (1987)

- Tennessee Williams: *A Streetcar Named Desire* (1951), *Baby Doll* (1956), *Cat on a Hot Tin Roof* (1958)

- Larry McMurtry: *Hud* (1963), *The Last Picture Show* (1971), ***Terms of Endearment*** **(1983)**

- Neil Simon: *The Odd Couple* (1968), *The Sunshine Boys* (1975), *California Suite* (1978)

- E.M. Forster: *A Passage to India* (1984), ***A Room with a View*** **(1985),** ***Howards End*** **(1992)**

- Stephen King: *Stand by Me* (1986), *The Shawshank Redemption* (1994), *The Green Mile* (1999)

As seen on the list above, two Oscar-winning screenplays owe their sources to E.M. Forster. Mario Puzo shares this distinction, with his novel *The Godfather* leading to two Best Adapted Screenplay wins: *The Godfather* (1972) and *The Godfather: Part II* (1974).

Like Puzo's *The Godfather* and Hammett's *The Thin Man*, three other writers had a single work adapted into two separate Oscar-nominated screenplays: George Bernard Shaw's play *Pygmalion* contributed to both the movie *Pygmalion* (1938) and the musical *My Fair Lady* (1964). Harry Segall's play *Heaven Can Wait* led to both *Here Comes Mr. Jordan* (1941) and *Heaven Can Wait* (1978). Richard

Linklater and Kim Krizan's *Before Sunrise* (1995) led to Academy-nominated sequels *Before Sunset* (2004) and *Before Midnight* (2013).

Puzo didn't just get lucky that some talented screenwriter got ahold of his novel: He took matters into his own hands, adapting his own work for the silver screen. These are two distinct talents, writing an original story and adapting one to work in movie format. Impressively, 14 individuals have done both of these tasks for different Oscar-nominated films, meaning they wrote but did not adapt one nominee, and adapted but didn't write another one (winners in bold):

Author	Wrote Source, but Didn't Adapt	Adapted, but Didn't Write Source
Maxwell Anderson	Anne of the Thousand Days (1969)	All Quiet on the Western Front (1930)
Viña Delmar	**Bad Girl (1931)**	The Awful Truth (1937)
Dashiell Hammett	The Thin Man (1934) After the Thin Man (1936) The Maltese Falcon (1941)	Watch on the Rhine (1943)
Moss Hart	You Can't Take It with You (1938)	Gentleman's Agreement (1947)
James Hilton	Goodbye, Mr. Chips (1939) Random Harvest (1942)	**Mrs. Miniver (1942)**
Lewis R. Foster	Mr. Smith Goes to Washington (1939)	The More the Merrier (1943)
Sally Benson	Meet Me in St. Louis (1944)	Anna and the King of Siam (1946)
Richard Brooks	Crossfire (1947)	Blackboard Jungle (1955) Cat on a Hot Tin Roof (1958) **Elmer Gantry (1960)** The Professionals (1966) In Cold Blood (1967)
Larry McMurtry	Hud (1963) **Terms of Endearment (1983)**	**Brokeback Mountain (2005)**
Ruggero Maccari	Scent of a Woman (1992)	Profumo di donna (1974)
Dino Risi	Scent of a Woman (1992)	Profumo di donna (1974)
Richard Price	Bloodbrothers (1978)	The Color of Money (1986)
Nick Hornby	About a Boy (2002)	An Education (2009) Brooklyn (2015)

Larry McMurtry is a particularly remarkable case. On top of being the only person with wins in both columns of the above chart, he is additionally the only person on that chart to also write both the source material and adaptation for yet another nominee, *The Last Picture Show* (1971).

All of the talented people in this chart didn't just get Oscar nominations for their work because of the type of source material they chose to adapt. They earned their nominations for their impressive adaptations of moving works, both fiction and nonfiction. Though if the frequency of nonfiction nominees keeps rising, perhaps it is best for Oscar-seeking writers to "comment on the world," as Abby Mann put it all those years ago.

Chapter 20. Best Actor

Do the BAFTAs have a British bias?

BRIGHTON:

That will do, Lawrence. Dreaming won't get you to Damascus, sir, but discipline will. Look, sir, Great Britain is a small country, much smaller than yours. Small population compared with some. It's small, but it's great. And why?

ALI:

Because it has guns.

BRIGHTON:

Because it has discipline.

–*Lawrence of Arabia* (1962): Nominated for 10 Oscars; won 7 (Best Picture, Best Director, Best Original Score, Best Art Direction: Color, Best Cinematography: Color, Best Film Editing, Best Sound)

I AM frequently asked, "What is the best predictor of the Oscars?" It's a straightforward-sounding question with a complex answer. Different predictors are useful for different categories. The Directors Guild, for instance, is an excellent forecaster for Best Picture and Best Director, but meaningless for Best Original Song.

I understand that's an unsatisfying answer. People want to know which awards show they should pay the most attention to in the lead-up to the Oscars. If I'm forced to pick one, my answer is always the same: the BAFTAs.

The British Academy of Film and Television Arts covers the majority of Oscar categories and has a decent track record across the board. But BAFTA's mission statement reveals a potential problem: "Our mission is to bring the very best work in film, games and television to public attention, and support the growth of creative talent in the UK and internationally."

UK and internationally. For an Oscar predictor, that's a loaded phrase. Does it mean that the BAFTAs judge all talent equally and just happen to be based in the United Kingdom? Or does it mean that the BAFTA voters lean towards rewarding British achievement? This is a key question to answer for the purpose of Oscar prediction. If the BAFTAs do in fact have a pro-British bias, we shouldn't give a British champion at the BAFTAs as much of a boost in Oscar chances as a non-British winner.

For this chapter, we will separate all films into whether or not they were produced (or partially produced) in the United Kingdom of Great Britain and Northern Ireland. We will separate all actors into whether or not they hold citizenship (or joint citizenship) in that same country.

Plenty of examples suggest a British bias, with a particularly extreme case arising in 1971. At the BAFTAs that year, all four acting awards went to British performers in British films for the first time ever:[1] Peter Finch and Glenda Jackson swept the lead acting honors for their roles in *Sunday Bloody Sunday*, an ahead-of-its-time depiction of a bisexual love triangle. Edward Fox and Margaret Leighton claimed the supporting trophies for *The Go-Between*, the story of a surreptitious turn-of-the-century affair.

The Oscars, however, disregarded that advice. Finch, Jackson, and Leighton were all nominated, but all lost. BAFTA winner Fox wasn't even nominated at the Oscars. Instead, the western side of the Atlantic chose Gene Hackman (*The French Connection*) and Jane Fonda (*Klute*) for the lead acting honors, and Ben Johnson and Cloris Leachman in *The Last Picture Show* for supporting awards.[2] An Oscar predictor that year would have been wise to look at the BAFTA results, consider them mere evidence of British bias, and ignore them entirely.

But plenty of examples go in the other direction. Consider the 2014 race for Best Actor. The BAFTAs picked Eddie Redmayne, an English actor playing Cambridge physicist Stephen Hawking in the UK production *The Theory of Everything*. Hard to get more quintessentially English than that. Many pundits and bettors overlooked Redmayne's BAFTA victory, instead predicting that Michael Keaton (*Birdman*) would claim the Oscar. They were wrong: Eddie Redmayne parlayed his London momentum straight to Los Angeles ballots, winning the Oscar over Keaton.

A few cases will really make you pause. Remember Glenda

---

1   This has happened only one time since, in 1988. John Cleese and Michael Palin won lead and supporting BAFTAs for *A Fish Called Wanda*. Maggie Smith won Best Actress for *The Lonely Passion of Judith Hearne*. Judi Dench won Best Supporting Actress for *A Handful of Dust*. None were nominated for Oscars.
2   Hackman, Johnson, and Leachman would all go on to win BAFTAs for these roles the following year, as the BAFTAs were on a different calendar from the Oscars at the time. But an Oscar predictor wouldn't have known that when making selections for the Academy Awards held in April 1972.

Jackson, whom I just cited as possible evidence of a BAFTA pro-British bias in 1971 a few paragraphs ago? Well, in 1970 and 1973, she acted in British films (*Women in Love* and *A Touch of Class*), lost the BAFTA both years, yet won the Academy Award for Best Actress each time. In those races, the Oscars were actually more pro-British than the BAFTAs were.

Instead of continuing to cherry-pick years, let's analyze all years to determine rigorously whether a BAFTA pro-British bias exists. Using data for the 50 years since 1968, when the BAFTAs introduced supporting acting categories and merged their British and non-British acting categories, let's examine all 1,000 Oscar acting nominees by BAFTA results:

BAFTA Result	Number of Oscar Nominees	Oscar Winning Percentage
No Nomination	534	12%
Nomination and Loss	320	18%
Win	146	53%

The first row of this chart says that 534 Oscar acting nominees did not receive a BAFTA nomination and 12% of them went on to win the Oscar anyway. Unsurprisingly, Oscar nominees who have BAFTA nominations on their résumé do better than those without BAFTA nominations (18% to 12%). And those who just won the BAFTA do a whole lot better (53% Oscar success rate) than those who lost.

Then the question becomes: How much better or worse do British contenders fare compared to what we'd expect based on the chart above? To make the analysis easier to follow, let's first see this in action for a specific example, the Best Actor nominees at the Oscars in 1989:

Actor	Film	BAFTA	Oscar Chance	Oscar
Daniel Day-Lewis	My Left Foot	Win	47%	Win
Kenneth Branagh	Henry V	Loss	15%	Loss
Tom Cruise	Born on the Fourth of July	Loss	15%	Loss
Robin Williams	Dead Poets Society	Loss	15%	Loss
Morgan Freeman	Driving Miss Daisy	N/A	10%	Loss

Grey shading indicates nominees from Great Britain. In each case, both the actor (Daniel Day-Lewis, Kenneth Branagh) and the film (*My Left Foot*, *Henry V*) are British.

As you can see, Daniel Day-Lewis won the BAFTA that year, Morgan Freeman wasn't even nominated, and the other three were all nominated but lost.[3] The column labeled "Oscar Chance" shows the nominee's chances at the Oscars based on his result at the BAFTAs. These don't precisely match the 12% / 18% / 53% from the earlier chart since I adjust each year to make the five percentages sum to 100%, which these five numbers do before rounding.

We now have to calculate how wrong we were for each nominee. For Daniel Day-Lewis, the BAFTA-based prediction was 47%, the probability he wins the Oscar before that final envelope is opened. After the envelope is opened, we learn that he did in fact win, which we'll call 100%. The difference is 100 − 47 = 53.

For Kenneth Branagh, Tom Cruise, and Robin Williams, we guessed 15% to win, but they actually scored 0% by losing the Oscar. In this case, the difference is 0 − 15 = -15. For Morgan Freeman, it's 0 − 10 = -10. We can perform the same calculation for all 1,000 acting nominees since 1968. Sometimes, the predictions will be too high. Other times, they will be too low. If they average out to being exactly

---

3   Cruise's BAFTA nomination came the following year due to differing calendars between the two awards organizations. As a result, these numbers don't precisely reflect what an Oscar predictor would have faced in March 1990, but we'll ignore that for now to keep our example simpler.

correct, then there is no bias. If, on average, our guesses for some group of nominees are too high or too low, then there may be a bias.

Here are the averages of those differences across years, within the groups of British actor in a British film (top-left), British actor in a non-British film (top-right), non-British actor in a British film (bottom-left), and non-British actor in a non-British film (bottom-right):

	Film	
	British	Not British
Actor: British	-10% (103 Nominees)	-2% (84 Nominees)
Actor: Not British	+5% (76 Nominees)	+1% (737 Nominees)

That top-left box, representing cases when *both* the actor and the film are British, shows a double-digit pro-British bias. On average, a BAFTA-based model of the Oscars makes predictions that are 10% too high on these doubly British nominees, so we need to subtract 10% from those predictions.

That's pretty substantial. The math is saying that the BAFTAs are too generous to these nominees, so we need to discount the BAFTA result. It means that a BAFTA-winning actor, who normally enjoys a 53% chance to win the Oscar, only has a 43% chance if he or she is both British and in a British film.

That said, in a five-actor race, 43% is still frequently good enough for first place. So even though we should discount the BAFTA result if it goes to a British actor in a British film, we should not throw it out entirely. The BAFTA winner is often the favorite – just not as strong a favorite as a non-British BAFTA winner is. After all, 11 men and women from the UK, acting in British films, have won the BAFTA as well as the Oscar:

Year	Category	Actor	Film
1969	Actress	Maggie Smith	The Prime of Miss Jean Brodie
1982	Actor	Ben Kingsley	Gandhi
1984	Actress (BAFTAs) Supporting Actress (Oscars)	Peggy Ashcroft	A Passage to India
1989	Actor	Daniel Day-Lewis	My Left Foot
1992	Actress	Emma Thompson	Howards End
1998	Supporting Actress	Judi Dench	Shakespeare in Love
2006	Actress	Helen Mirren	The Queen
2010	Actor	Colin Firth	The King's Speech
2014	Actor	Eddie Redmayne	The Theory of Everything
2014	Actress	Julianne Moore	Still Alice
2017	Actor	Gary Oldman	Darkest Hour

Yes, the British are somewhat inclined to honor their own – who can blame them? – but it's undeniable that the island has produced decades of remarkable performances, many of which have been honored both in the UK and in Hollywood.

Chapter 21. Best Actress

Does playing a historical figure help win an Oscar?

> **WILLIAM:**
> Don't you have any regular friends?
>
> **PENNY:**
> Famous people are just more interesting.
>
> –*Almost Famous* (2000): Nominated for 4 Oscars; won 1 (Best Original Screenplay)

WHEN HILLARY Swank won Best Actress for *Boys Don't Cry* (1999), she concluded her speech with a moving tribute to the film's subject: "And last, but certainly not least, I want to thank Brandon Teena for being such an inspiration to us all. His legacy lives on through our movie to remind us to always be ourselves, to follow our hearts, to not conform. I pray for the day when we not only accept our differences, but we actually celebrate our diversity. Thank you very much."

Conversely, A.O. Scott wrote in his *New York Times* review of *Lincoln* (2012) that, "The most famous and challenging beard of them all sits on the chin of Daniel Day-Lewis, who eases into a role of epic difficulty as if it were a coat he had been wearing for years." In one sentence, Scott describes the role as both "challenging" and "of epic difficulty." What makes the part of Abraham Lincoln so difficult? Later in the review, Scott added, "Mr. Day-Lewis, for his part, must convey both the human particularity and the greatness of a man who is among the most familiar and the most enigmatic of American leaders."

Swank said that the life story of hate crime victim Brandon Teena inspired her along with the rest of the cast and crew when making this film. Surely, this inspiration can make an actor's job easier. But it's hard to argue with Scott's notion that portraying a well-known historical figure can be more challenging because audiences arrive with an expectation of how that person acts, talks, and moves.

Though we can't use data to measure whether real-life roles are tougher to tackle, we can calculate whether actors and actresses who play historical figures have won Oscars more or less frequently than those who play fictional characters.

But before we can do that, we have the difficult task of deciding which roles are based on real people and which are based on imagination. There is no perfectly objective way to do this, but I did my best, going through all 1,708 acting nominees in all four categories (male and female, lead and supporting) and applying the following standards to each one:

1. I defined playing a historical figure as portraying any character based on a real-life person. That includes not only famous people – like Ben Kingsley's Oscar-winning portrayal of Mohandas Gandhi in *Gandhi* (1982) or Greer Garson's Oscar-nominated performance as Marie Curie in *Madame Curie* (1943) – but also lesser-known individuals: Albert

Finney as Edward Masry in *Erin Brockovich* (2000) and Lupita Nyong'o as Patsey in *12 Years a Slave* (2013) both count as well. This definition is purposefully broad because it would be too subjective to try to draw a line between "famous" and "not famous."

2. I also counted as historical figures characters based on real people whose names or details were changed. There was never a governor of Louisiana named Willie Stark, but it's widely known that the fictional politician played by Oscar winner Broderick Crawford in *All the King's Men* (1949) was based on real-life governor Huey Long.

3. I drew the line at composite characters. In *Zero Dark Thirty* (2012), Jessica Chastain plays a CIA agent named Maya. Her role was a combination of multiple real-life CIA agents, so that does not count as a historical figure for this chapter. To be included in the group of actors who based their performances on real people, the part had to be based on just one person.

4. An intriguing corner case: In *Adaptation.* (2002), Nicholas Cage played a fictionalized version of Charlie Kaufman, the author of the film's script. By my definition, even a fictionalized version of a real person counts. But thanks to some *Parent Trap*-esque technology, Cage also plays Charlie's brother Donald, who shares screenwriting credit and even received an Oscar nomination for writing the film. Just one problem: Donald is completely fictional, a bizarre alter ego invented by Charlie Kaufman for this script. So, this is the only nomination to which I assigned a 0.5, where 0 means fictional and 1 means real.

In some years, an actor or actress was nominated for multiple roles. At the inaugural ceremony, before the rules changed to the modern structure, Emil Jannings, Richard Barthelmess, and Janet Gaynor received nominations for multiple films. A few others have been nominated for multiple roles within the same film: José Ferrer played both Henri de Toulouse-Lautrec and his father in *Moulin Rouge* (1952). Lee Marvin won Best Actor for playing a pair of brothers – one a drunken hero, the other a villainous outlaw – in *Cat Ballou* (1965). Klaus Maria Brandauer played a pair of identical twins in *Out of Africa* (1985). Peter Sellers played three parts in *Dr. Strangelove* (1964). Fortunately, none of the cases mentioned in this paragraph were difficult to classify because all of the roles were either fiction or nonfiction for each of these performers.

Let's begin by looking at the breakdown by category:

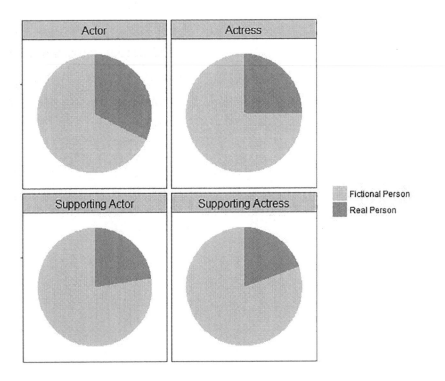

Strangely, even though Best Supporting Actress is the least likely of the four acting categories to have a nominee based on a real person, it also played host to the longest streak of consecutive winning historical roles. From 1998-2002, all five winners played actual people: Judi Dench as Queen Elizabeth I in *Shakespeare in Love* (1998), Angelina Jolie as sociopath Lisa Rowe in *Girl, Interrupted* (1999), Marcia Gay Harden as painter Lee Krasner in *Pollock* (2000), Jennifer Connelly as mathematician John Nash's wife Alicia in *A Beautiful Mind* (2001), and Catherine Zeta-Jones as murderer Velma Kelly (based on nightclub singer Belva Gaertner) in *Chicago* (2002).

In the other direction, the longest streak for fictional champions in one category is 15, thanks to a 1961-1975 run in the Best Supporting Actor race. Every winner from George Chakiris as a gang leader in *West Side Story* (1961) to George Burns as an aging comic in *The Sunshine Boys* (1975) played a made-up character.

In 28 years, no winner in any of the four acting categories played a historical role. In only one year – 2002 – did all four winners portray real-life individuals: Adrien Brody as Holocaust survivor Władysław Szpilman in *The Pianist*, Nicole Kidman as author Virginia Woolf in *The Hours*, Chris Cooper as poacher John Laroche in *Adaptation.*, and Catherine Zeta-Jones as Velma Kelly in *Chicago*.

The fact that this didn't occur until 2002 isn't too surprising, given that the Oscars have witnessed a marked increase in the frequency of historical nominees:

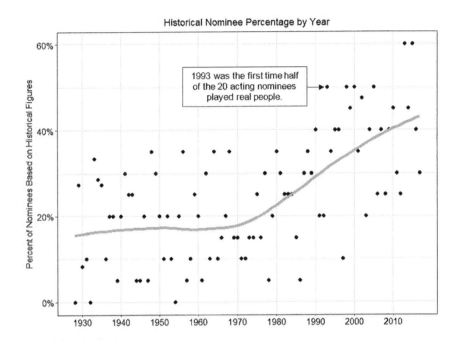

Historical Nominee Percentage by Year

The grey curve[1] shows the trend over time, which is rising over the decades. Sure enough, the only two times in history that more than half of all acting nominees were nonfiction occurred quite recently, in 2013 and 2015, at 60% apiece (those two black points in the top right of the graph).

To answer the original question, asking how much playing a historical figure helps or hurts, it will be useful for us to break this dataset into groups of years according to how many nominated parts in each year are based on real people and how many are created from scratch.

We will disregard years in which all five nominees in a category played fictional characters, because those years don't tell us anything about whether fictional nominees have an advantage or disadvantage. There has never been a category in which all five nominees played real

---

1   For those who are interested, this is known in statistics as a *LOESS curve*, which is a curve that tries to get as close as possible to the nearby points in each region of the plot.

people. First up: The set of years in which exactly one nominee in a category played a historical figure and the other four played imaginary roles:

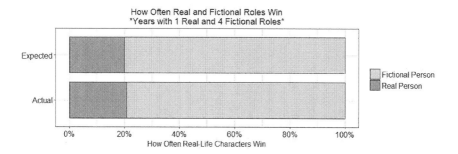

The top bar, labeled "Expected," shows how often we would *expect* the real and fictional nominees to win, if there were no advantage or disadvantage based on which category a nominee falls in. The darker bar, representing real nominees, is at 20%, because with one out of five nominees playing a real character, we would expect the real character to win one-fifth of the time.

The bottom bar, labeled "Actual," shows how often real and fictional nominees *actually* won in these years. Since the lengths of the two darker bars are nearly even, these cases do not provide any evidence of an advantage or disadvantage to playing a real person. Let's move on to years with two real nominees and three fictional ones:

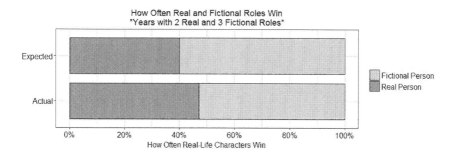

In this chart, the expectation is that real nominees will win 40% of the time and fictional ones will win 60%, since two out of five nominees played real-life characters. When we look at actual history, however, 47.3% of winners in this group are real, suggesting an advantage for real nominees. Next up: years when three of the nominees are real and two are fictional.

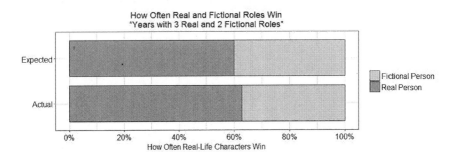

It's a similar story – the bar for real nominees is slightly longer in actuality than in expectation, implying a small advantage for real nominees. Finally, we look at years when four real characters and one fictional one went head-to-head at the Oscars:

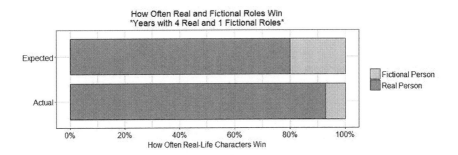

As in the last two groups, there is also an advantage in these years to portraying a true person. These gaps are not giant, but over 90 years of Oscar history, they do show a modest pattern working against actors and actresses who breathe life into a character without real-life inspiration.

So, for all you actors out there: Many considerations go into choosing which role is most likely to win you that Oscar, from whether you fit the part, to which director is attached to the project, and so on. But all else being equal, choosing a role modeled on an historical figure is a bit more likely to put your performance in the Oscar history books.

Chapter 22. Best Director

Does a Best Director win increase Best Picture odds?

> MENDEZ:
>
> Can you teach somebody to be a director in a day?
>
> CHAMBERS:
>
> You can teach a rhesus monkey to be a director in a day.
>
> –*Argo* (2012): Nominated for 7 Oscars; won 3 (Best Picture, Best Adapted Screenplay, Best Film Editing)

SAY THAT I just paused *The Usual Suspects* (1995) the first time the name Keyser Söze is mentioned to go grab some popcorn. While the popcorn is in the microwave, I ask you – a first-time viewer of this riveting film – to predict which character is Keyser Söze. You'd have a pretty tough time guessing with much confidence. Five criminals are in the police lineup at the beginning of the film, and all of them seem to have a roughly equal probability of being the arch villain.

The microwave beeps, I pour the popcorn into a bowl, and we resume the movie. An hour later, with just five minutes remaining in the film, I accidentally sit on the remote, pausing the movie yet again. I hope you'll forgive me for interrupting one of the most stunning conclusions in the history of storytelling. While I'm trying to retrieve the remote from underneath the couch cushions, care to venture another guess as to the real identity of Keyser Söze? I bet your prediction will be a lot more accurate this time, and you'll have a lot more confidence in your pick than you did an hour ago.

This is an apt metaphor for how people talk about predicting Best Picture – before and after the Best Director award is announced. Before the director envelope is opened, we're all like the first-time viewer at the moment Keyser Söze's name is initially heard. We have at least five Best Picture options, and all of them seem like plausible guesses. After Best Director is handed out, we're like the viewer near the end of the movie, when we feel fairly confident guessing the ending of who wins Best Picture. Of course, like the viewer waiting for the twist at the movie's very last moment, we might still be wrong.

Is conventional wisdom right? Does winning Best Director increase the chances that the same film wins Best Picture?

Intuitively, we expect the answer to be yes. A director is the glue that ties a film together. Certainly, some directors are more influential than others: Orson Welles conceived of and executed *Citizen Kane* (1941) like no one else could have, while Michael Curtiz was part of a large team that came together to create *Casablanca* (1942). But in both cases, the director's influence played a major role in making the film so outstanding. The conventional wisdom that winning Best Director helps a film land Best Picture makes a lot of sense.

But if it does help, then by how much? For instance, based on a model I will explain shortly, the neurotic comedy *Annie Hall* (1977) had a 65% chance to win Best Picture as it entered the Best Director race. Then, Woody Allen won Best Director for the movie. How high are *Annie Hall*'s Best Picture chances now? 70%? 80%? 90%?

Instead of guessing, let's do some math.

In the 90 years of Oscar history, the Best Director winner also claimed Best Picture 65 times (that is, 72%). The tradition started off slowly – in the first five years, only the tragic World War I story *All Quiet on the Western Front* (1930) took both honors. Then from 1933 to 2011, the rate jumped to 78%. Since then, what's old is new again, as four of the last six years saw different Best Picture and Best Director winners, just like those first five years. *Birdman* (2014), the movie that appears to be one long tracking shot, and *The Shape of Water* (2017), the 1960s adventure about a woman falling in love with an amphibian, are the only two during the most recent stretch to claim both trophies.

Does that mean that all Best Director winners have a 72% shot to win Best Picture? Well, if all we knew was which film won Best Director, that would be a reasonable estimate for Best Picture. But we know more than that – lots more, in fact. We know the nominations in every category, the winners of all the previous categories that night, and the results of all the pre-Oscars awards.

To take all of that information into account, I built a pair of models that I'll use throughout this chapter:

1. **The first model predicts Best Director.** It uses all data that's available every year from 1948 to the present, since 1948 was the first year of the Directors Guild Awards, and then assigns weights to this data based on how closely it has corresponded with the Best Director result. Namely, the model includes:

- The results of previous awards shows: the Directors Guild of America (DGA) Awards, British Academy of Film and Television Arts (BAFTA) Awards, and Golden Globe Awards.

- The nominees and winners of previously announced categories: Best Cinematography, Best Costume Design, Best Film Editing, Best Original Score, Best Production Design, Best Sound Mixing, the two screenplay awards, and the two supporting actor awards.

- The nominees (but not winners) of categories that are typically announced *after* Best Director: Best Actor, Best Actress, and Best Picture.

2. **The second model predicts Best Picture.** It uses just two pieces of information: the likelihood that the film wins Best Director (based on the previous model, before we know the Best Director result), and the actual result of the Best Director race. Naturally, we expect movies with a higher initial chance of winning Best Director to also be stronger contenders for Best Picture. We also expect movies that actually win Best Director to have better Best Picture chances. The question is how much better.

Let's look at the predictions from these models for 2003. *The Lord of the Rings: The Return of the King* entered the night as the heavy favorite, not only in a number of technical categories that it obviously deserved to win, but also for Best Director and Best Picture. Then, it went 9-for-9 to begin its night. After that remarkable run, the model gave the movie an 88.5% chance to win Best Director, and a 77.1% chance to win Best Picture, before the Best Director announcement.

Film	Best Director Odds	Best Picture Odds	
		Pre-Best Director	Post-Best Director
The Lord of the Rings III	88.5%	77.1%	79.1%
Master and Commander	5.8%	6.5%	5.7%
Lost in Translation	2.9%	5.7%	5.2%
Mystic River	2.0%	5.5%	5.1%
Seabiscuit	0.0%	5.0%	4.8%

After Peter Jackson took home his expected Best Director honor, making the Middle-earth film 10-for-10, its Best Picture odds increased from 77.1% to 79.1%. As predicted, the movie went on to win Best Picture, wrapping up an astounding 11-for-11 night, breaking the record of most wins without a loss previously set at 9-for-9 by *Gigi* (1958) and *The Last Emperor* (1987).

So, looking at the above chart, the Best Director win is worth … only 2.0%? After all, *Lord of the Rings* jumped from 77.1% to 79.1%. Seems a bit small, right? Before we explore why, let's perform the same analysis for the previous year, 2002.

According to my models, *Chicago* entered the Best Director race with a 78.0% chance to win Best Director and a 70.4% chance to win Best Picture. And then, just as the night was winding towards its conclusion, a dramatic upset occurred: *The Pianist* won Best Director.

Film	Best Director Odds	Best Picture Odds	
		Pre-Best Director	Post-Best Director
Chicago	78.0%	70.4%	46.5%
The Pianist	12.0%	9.7%	31.6%
Gangs of New York	5.8%	7.4%	7.9%
The Hours	3.1%	6.6%	7.3%
The Lord of the Rings II	0.0%	5.8%	6.7%

Hold on a second! When *The Pianist* won Best Director, its Best Picture chances jumped from 9.7% to 31.6%. That's a 21.9% difference, which is a whole lot more than the 2.0% improvement that *The Lord of the Rings: The Return of the King* got after it won Best Director.

Had Twitter been invented this early, I'm sure a number of real-time commentators would have instantly predicted *The Pianist* to take Best Picture once it won Best Director. But math doesn't overreact. Yes, *Chicago*'s chances plummeted from 70.4% to 46.5%, and what was a blowout race only moments earlier became a tight battle. But *Chicago* still led *The Pianist*, 46.5% to 31.6%, and sure enough, the jazzy musical did in fact win the evening's final prize, becoming the first musical to win Best Picture since *Oliver!* (1968).

So why are these years so different in terms of how much Best Director affected the predictions for the Best Picture race? A couple of reasons:

First, *The Lord of the Rings: The Return of the King* simply didn't have as much room to improve. It was already at 77.1% to begin with, which is awfully high – by this model, no film has ever been higher than 80.7% to win Best Picture, a record held by the classic tale of man versus desert, *Lawrence of Arabia* (1962). If math is already highly confident in something happening, no additional information is ever going to move the numbers all that much. *The Pianist*, on the other hand, had lots of room to improve, starting out at just 9.7%, coming in a bit below *Traffic* (2000) for the awkward honor of worst Best Picture odds among Best Director winners.

Second, in 2002 the Best Director announcement provided us a lot of new information. In 2003, it did not. Let me explain: *The Pianist* had a 12.0% chance to win Best Director, so when it actually did win that category, it was a huge upset. When the math learned that *The Pianist* won Best Director, it learned something it definitely didn't know before.

The Lord of the Rings III, in 2003, had an 88.5% chance to win Best Director. That high percentage was already baked into its Best

Picture chances. So when it actually won Best Director, we didn't gain much new information that the math didn't already have. It already basically knew who was going to win Best Director, with 88.5% confidence, so why change the Best Picture prediction once the Best Director announcement is made?

Now we have our answer for how much the Best Director result affected the Best Picture races in 2002 and 2003. But it would be nicer to have a general rule of thumb for all years. So, let's put the results of these models into a graph.

- The middle, dashed curve shows the relationship between the chance that a movie wins Best Director and the initial chance that it wins Best Picture. Unsurprisingly, the line slopes upwards: Movies that are more likely to win Best Director are also more likely to win Best Picture.

- The top curve deals with films that already *won* Best Director. It shows the relationship between how likely the film was initially to win Best Director (before it was announced as the winner) and how likely the film is to win Best Picture.

- The bottom curve deals with films that already *lost* Best Director. It shows the relationship between how likely the film was initially to win Best Director (before it lost) and how likely the film is to win Best Picture.

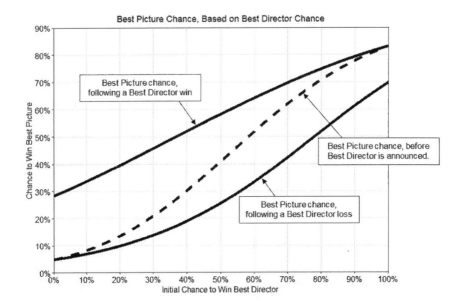

Note that once a movie wins Best Director, its Best Picture chances can only go up. Once a movie loses Best Director, its Best Picture chances always fall.

Instead of going through the results of every year, one by one, and examining how much the Best Director result affected the Best Picture race (like I did for 2002 and 2003), we can instead use a handy trick for describing the Best Director effect for most years all at once: using the Directors Guild winner as a proxy for the Best Director favorite.

When I built the model to predict Best Director, I confirmed a fact that will be obvious to followers of awards season: The best predictor of Best Director, by far, is the Directors Guild of America. The Directors Guild, after all, is filled entirely with accomplished directors, many of whom have been nominated for or won Oscars themselves. They tend to know a thing or two about what qualifies as good directing. Plus, many members double as Academy voters and often vote for the same film for both awards.

Since 1950, when the Directors Guild and the Oscars got on the same calendar, the Directors Guild has only been wrong about Best Director seven times:

Year	Directors Guild	Best Director Oscar
1968	The Lion in Winter	Oliver!
1972	The Godfather	Cabaret
1985	The Color Purple	Out of Africa
1995	Apollo 13	Braveheart
2000	Crouching Tiger, Hidden Dragon	Traffic
2002	Chicago	The Pianist
2012	Argo	Life of Pi

That's a remarkably strong track record. Of these seven misses, only three of the Directors Guild winners weren't even nominated for the Best Director Oscar: *The Color Purple* (1985), *Apollo 13* (1995), and *Argo* (2012). Even more impressively, there has never been a single year in which the Directors Guild champion wasn't nominated for Best Picture.

So, in all years aside from 1985, 1995, and 2012, the Directors Guild winner received both a Best Director and a Best Picture nomination. This means there was a heavy favorite for Best Director entering the night, since the Directors Guild corresponds so well with that category, and that favorite was also competing for Best Picture. We can look at what happens to an average Directors Guild winner's Oscar chances, and that will give us a fairly good idea of what takes place in a typical year.

Of the 65 Best Director nominees to win the Directors Guild Award since 1950, the average Best Director probability by the model I introduced at the beginning of the chapter is 81.8%. I then used my graph from earlier in the chapter to see what happens to a movie with an 81.8% chance to win Best Director. It tells me that that the movie (an average Directors Guild winner) will initially have a 72.2% chance to win Best Picture. If it goes on to win Best Director, that chance will jump to 75.6% for Best Picture. If it instead goes on to lose Best Director, it will fall to a 53.6% chance to win Best Picture.

So, if the Directors Guild winner takes Best Director, you should barely change your Best Picture prediction at all. The favorite only jumps 3.4%, from 72.2% to 75.6%, which is not a huge difference. We already knew the movie was likely to win Best Director, thanks to the Directors Guild, so actually winning Best Director doesn't tell us much.

But if the Directors Guild winner loses Best Director, that's a horse of a different color. That movie just lost almost 19 percentage points, from 72.2% to 53.6%. Approximately, that 19% gets reassigned to the movie that just pulled off the Best Director upset.

If the Best Picture race was previously a 37-point spread or closer, it's a tight enough race that subtracting 19% from the frontrunner and adding 19% to the film in second would change the expected champion. If there's a heavy favorite coming in – a movie that has more than a 37% lead on second place – it's still going to be the favorite no matter who wins Best Director.

During the final commercial break before the Best Picture reveal, you can now sound pretty smart at your Oscar party. If the Directors Guild winner just won Best Director, you can say, "I'm still just as confident that movie is going to take Best Picture."

But if there was an upset for Best Director, tread carefully. If there was a heavy Best Picture favorite coming in, you should say, "I still think the Best Picture favorite is going to win, even though it lost Best Director a moment ago." But if the Best Picture race started out tight, instead go with, "This changes everything. The movie that just pulled off a Best Director upset is going to ride this momentum to a Best Picture victory."

Then the commercial break ends, and the drama begins.

Chapter 23. Best Picture

The biggest upsets in Oscar history

> JOHN PRENTICE:
> You may be in for the biggest shock of
> your young life.
>
> —*Guess Who's Coming to Dinner* (1967): Nominated for 10 Oscars; won 2 (Best Actress, Best Original Screenplay)

THE MOST shocking moment in Oscar history, without a doubt, was the conclusion to the 89th Academy Awards. For those reading this book shortly after its publication, I'm confident the stunning turn of events is still fresh in your minds. For readers who happen to stumble upon this book decades into the future, a refresher:

The final award of the night was introduced by Hollywood legends Warren Beatty (a 14-time nominee, who won Best Director for *Reds* in 1981) and Faye Dunaway (a 3-time nominee, who won Best Actress for *Network* in 1976). It was, after all, the 50th anniversary of their classic *Bonnie and Clyde* (1967), whose memorably bloody ending set a new standard for how much violence is acceptable on-screen. The bank robbers' tale won Oscars for Best

Supporting Actress and Best Cinematography, and was nominated for eight others including Best Picture.

Beatty opens the red envelope, and hesitates, and hesitates. Eventually, he hands the envelope over to Dunaway, who promptly announces that *La La Land* won Best Picture. Only … it didn't. A few speeches in, a stagehand informs the *La La Land* brass that there has been an error. Beatty was handed the wrong envelope. One of that film's producers then delivers the stunning news to the Dolby Theatre and the television audience: It's *Moonlight*.

This was not the first such snafu in Academy history. At the 1933 Oscars, host Will Rogers announced Best Director by stating, "Come on up and get it, Frank!" Frank Capra, nominated for directing the darling rags-to-riches tale *Lady for a Day*, leapt towards the podium. But the winner was Frank Lloyd, director of the lengthy British drama *Cavalcade*. Capra had every right to be upset – in hindsight, *Lady for a Day* was the better-directed film. The story of a whole town coming together to support a poor old lady's charade is as powerful now as it was then, while *Cavalcade* manages to make a film covering 34 years of British history feel 34 hours long.

Thirty years later, Sammy Davis Jr. was handed the Best Original Score envelope, declaring the unusual British comedy *Tom Jones* the winner, when he was supposed to announce that Best Adapted Score belonged to the Parisian romance *Irma la Douce*.

But neither of those events endured nearly as heavily in the Academy's collective memory as I predict the *Moonlight* mix-up will. It's not just that the 89th ceremony had a wider audience or that the error occurred on the biggest category of them all. It's also due to the magnitude of the shock. Nearly every prognosticator was picking *La La Land*, so the big reveal that it did not in fact win Best Picture was nearly as stunning as the manner in which the events unfolded on stage.

Some writers speculated that it was the largest Oscar upset of all time – a reasonable proposition. But with a little math, we can do better than mere guessing.

For this chapter, I ran a variation of my normal Best Picture model. The one I publish each year only includes data accumulated before the Oscars ceremony, such as other awards shows and which categories a film is nominated in. The one I use here will also include which other Oscar categories a film *wins*, since the goal is to determine which upsets were the most shocking the moment the envelope is opened.

I used data from every year since 1948, the first year we have Directors Guild, BAFTA, and Writers Guild results, which are three of the biggest players in the Oscar predictions game. From this model, it turns out that *Moonlight* defeating *La La Land* is in fact the fourth-biggest upset in Oscar history, measured by how far the actual winner entered the night behind the projected winner. Let's break down the rest of the top seven:

7. *Braveheart* (18.9% chance to win Best Picture, based on my model) defeats *Apollo 13* (50.3% chance to win), 1995
Mel Gibson's thirteenth-century Scottish war movie was trying to pull off an upset both on- and off-screen. In the film itself, William Wallace leads a band of ragtag Scotsmen into battle against the mighty British army in hopes of earning freedom for him and his countrymen. In Hollywood, the movie sought to knock off *Apollo 13*, Ron Howard's dramatic retelling of the doomed spacecraft and the heroic astronauts and NASA engineers who brought the crew safely down to Earth.

Just as King Edward I's troops had superior firepower, *Apollo 13*'s Oscar résumé had the stronger listing, with wins from the Directors Guild and Screen Actors Guild. *Braveheart* wasn't even nominated at the Screen Actors Guild awards, the first year a Best Cast honor was bestowed, and was the only Best Picture winner until *The Shape of Water* (2017) to top the Oscars without a SAG nomination.

They may take our guild awards, but they'll never take our Oscar.

6. *Driving Miss Daisy* (26.6%) defeats *Born on the Fourth of July* (61.2%), 1989

This was the first of Billy Crystal's 9 times to host the Oscars, second only to Bob Hope's 18. In his leadoff attempt, he delivered one of his most memorable zingers to the Academy, saying that *Driving Miss Daisy* "apparently directed itself." Indeed, the Academy overlooked Bruce Beresford's directing on the slow-paced Atlanta picture – a slight often considered the death knell for Best Picture hopefuls.

Born on the Fourth of July, on the other hand, seemed perfectly primed. Vietnam veteran Oliver Stone had just won Best Picture three years earlier for another Vietnam film, *Platoon*, which in many ways is the thematic prequel to *Born on the Fourth of July*. The 1986 film addresses the horrors of the war itself, while the 1989 one deals movingly with the tribulations of returning home.

Nevertheless, *Driving Miss Daisy* must have done a good job directing itself, because it pulled off the upset for Best Picture.

5. *The Greatest Show on Earth* (15.0%) defeats *The Quiet Man* (49.0%), 1952

Minutes before Best Picture was announced, viewers of the first televised Oscar broadcast were treated to quite the Best Director race, as five of the biggest names in Hollywood vied for the honor: Cecil B.

DeMille (*The Greatest Show on Earth*), John Ford (*The Quiet Man*), John Huston (*Moulin Rouge*), Joseph L. Mankiewicz (*5 Fingers*), and Fred Zinneman (*High Noon*). John Ford won the directing honor yet again – his fourth win in that category following *The Informer* (1935), *The Grapes of Wrath* (1940), and *How Green Was My Valley* (1941) – a record that still stands to this day.

Many critics at the time believed *High Noon*'s story of a lone marshal (Gary Cooper) taking on a gang of outlaws to be the favorite. But with Ford's win for *The Quiet Man*, a John Wayne and Maureen O'Hara romance most memorable for its lovely Irish vistas, the math says Ford cemented his film's frontrunner status for Best Picture. And yet, the Academy announced its decision to finally honor another seminal director, Cecil B. DeMille, with the Best Picture statue. DeMille's circus drama may not have stood the test of time, but it earned a piece of history by securing the night's final trophy.

4. *Moonlight* (21.5%) defeats *La La Land* (62.3%), 2016

As described above, this upset is unique: It's more memorable for the way that it happened than the fact that it happened. But it's an upset nonetheless. Nearly all major predictors went against *Moonlight*. Barry Jenkins' movie is the only film in history to win Best Picture after losing the top awards from the Directors Guild, the BAFTAs, the Producers Guild, the Screen Actors Guild, and the American Cinema Editors.

3. *Spotlight* (19.1%) defeats *The Revenant* (63.4%), 2015

In the vein of *All the President's Men* (1976), *Spotlight* tells the true story of *Boston Globe* reporters uncovering widespread sexual abuse in the Catholic Church. The film won the Screen Actors Guild Award for Best Cast, but was largely shut out at the other major awards shows. Prior to the Best Picture announcement, *Spotlight* had secured but one Oscar for Best Original Screenplay.

The Revenant is one of the bloodiest Best Picture nominees of the last decade, as Oscar winner Leonardo DiCaprio plays a nineteenth-century frontiersman who is nearly killed and left to die, and then spends the rest of the movie seeking vengeance. In addition to DiCaprio's numerous awards, the film won Best Director as well as top honors from the Directors Guild, the Golden Globes, and the BAFTAs.

And yet, with the spotlight on the Dolby Theatre stage, *Spotlight* won Best Picture, becoming the first Best Picture winner since *The Greatest Show on Earth* (1952) to come away with only two trophies.

2. *An American in Paris* (23.2%) defeats *A Place in the Sun* (71.4%), 1951

Though less remembered today, *A Place in the Sun* earned significant critical acclaim back in 1951. A film luminary no less than Charlie Chaplin declared it "the greatest movie ever made about America." Based on the classic American novel *An American Tragedy*, it tells the story of a man who impregnates one woman but falls in love with another, and the disastrous consequences that follow when he tries to keep his two lives separate.

A Place in the Sun could hardly be more different from *An American in Paris*, a film heavy on ballet and light on plot. *An American in Paris* features oodles of splashy color at a time when only one color film had ever won Best Picture – *Gone with the Wind*, a dozen years prior. So, the Academy was presented with a choice between a depressing novel adaptation and a series of George and Ira Gershwin songs.

In the end, George Stevens took Best Director for *A Place in the Sun*, but the last laugh went to *An American in Paris* when it won Best Picture.

1. *Crash* (23.5%) defeats *Brokeback Mountain* (73.0%), 2005
Unlike the winners in this chapter, I doubt this race topping the list will surprise anyone. *Crash* is now widely considered to be one of the worst Best Picture winners of all time, and understandably so. It tried to present the issue of racism with more nuance than previous films, but ended up lumping together a hundred actors into an ensemble story that was poorly organized from start to finish. Even its director, Paul Haggis, told HitFix in 2015, "Was it the best film of the year? I don't think so."

Brokeback Mountain told the tragic tale of two men in love in a society that refused to accept them. It's a less complex film than *Crash*, but Ang Lee handled it well, guiding the audience's emotions with every movement of his camera. Lee won his second Directors Guild honor, following the one for *Crouching Tiger, Hidden Dragon* (2000).

However, at the time, there was at least some support for *Crash*. Roger Ebert declared it the best movie of the year, and the Academy awarded it honors for its screenplay and editing, along with Best Picture.

Was 1998 an upset?
Shakespeare in Love's win over *Saving Private Ryan* is widely considered to be one of the biggest upsets of all time, but had Oscarmetrics been around in 1998, it wouldn't have been nearly so high on *Ryan*'s chances. *Shakespeare in Love* won top honors at the BAFTAs and the Screen Actors Guild to go along with its Golden Globe for Best Comedy/Musical and other awards. It led the Oscar nominations with 13 entries, and before Best Picture was announced, it had already won 6 trophies, the most of the evening. *Saving Private Ryan* was certainly a strong contender in its own right, with its Directors Guild win, Golden Globe for Best Drama, and Best Director award. But this was a much closer race than people realized.

• • •

A review of the greatest upsets in Oscar history wouldn't be complete without including the first 20 years of film awards. While we can't rank them mathematically due to a lack of data, a couple deserve mention:

Wings defeats *7th Heaven*, 1928

In the run-up to the first year, there would have been no way to mathematically determine which films were favored. There was no prior history to suggest that nominations in certain categories would lead to a Best Picture win, and there were no other film awards to point the way in advance.

That said, consider the résumés of two of the contenders that year. *7th Heaven*, a moving silent film about how love always finds a way to survive, helped put Fox on the map in Hollywood. It was nominated for Best Picture and Best Art Direction, and won Best Director of a Drama, Best Actress,[1] and Best Adapted Screenplay. *Wings*, a technical wonder thanks to its incredible scenes of World War I aircraft combat, was nominated for Best Picture and Best Engineering Effects, but nothing else.

Even without historical precedent, it's a reasonable assumption that Best Art Direction plus Best Director plus Best Actress plus Best Adapted Screenplay should be given more weight than Best Engineering Effects. But *Wings* took flight and claimed the inaugural award.

---

1   Janet Gaynor's Best Actress win was for three films simultaneously – *7th Heaven*, *Street Angel*, and *Sunrise* – due to different rules in the initial year.

Grand Hotel defeats *The Champ*, 1932

Grand Hotel, a loosely patched together series of vignettes about unrelated people in the same fancy hotel, remains to this day the only Best Picture to not be nominated for anything else.

The Champ, King Vidor's tear-jerking piece about an alcoholic boxer trying to provide his son with a good life, had nominations for Best Picture, Best Director, Best Actor, and Best Story, winning the latter two categories. That Best Actor win went to Wallace Beery, who played the title character. But it's the Champ's son, portrayed by nine-year-old Jackie Cooper, who steals the show. In my mind, it's the strongest child acting performance of all time, with all due respect to Haley Joel Osment in *The Sixth Sense* (1999) and Lindsay Lohan in *The Parent Trap* (1998). But none of this was enough for *The Champ* to take the top prize away from *Grand Hotel*.

Was 1941 an upset?

Finally, film buffs may be puzzled at the absence of one very memorable year from this list: 1941. That was the year that *How Green Was My Valley* defeated *Citizen Kane*. The former is a forgettable tale of striking Welsh miners; the latter is now widely considered to be among the twentieth century's greatest pieces of art, largely for the remarkable number of filmmaking techniques that were invented for or popularized by Orson Welles' innovative work.

But this is not a list of the Academy's mistakes. This is a list of upsets, meaning results that were probabilistically surprising at the time of the ceremony. It does not allow for the benefit of hindsight.

The only relevant precursor awards in 1941 were the New York Film Critics Circle (NYFCC) and the National Board of Review (NBR), but neither could be trusted. To that point, NYFCC had only predicted one Best Picture winner in six tries (1937's *The Life of Emile Zola*), and NBR was an even worse one-for-nine (matching the Oscars on 1934's *It Happened One Night*). Yes, *Citizen Kane* had Oscar nominations for directing, screenplay, acting, score, sound, art direction, cinematography, and editing. But so did *How Green Was My Valley*, matching its more famous foe nomination for nomination plus one more: *How Green Was My Valley* garnered two acting nominations to *Kane*'s one.

Historians have blamed everything from studio extras voting against *Citizen Kane* en masse to William Randolph Hearst (the obvious inspiration for the semi-fictional Charles Foster Kane) running a smear campaign against the film. But the fact of the matter is, 1941 was a close race at the time. As we'll see in the next chapter, it is only with the benefit of hindsight that *Citizen Kane* rises to the top.

Chapter 24. After-Party

What should have won?

> **KANE:**
> That's a mistake that will be corrected one of these days.
>
> –*Citizen Kane* (1941): Nominated for 9 Oscars; won 1 (Best Original Screenplay)

THE MOST frequent criticism of my work predicting and analyzing the Oscars is some variant of the following: "You can't use *math* to figure out which movie is the best!" That's an entirely true statement, but it's not a valid criticism of what I do. Predicting the Oscars and determining which movie is the best are two totally different undertakings. What I'm trying to do is estimate how a group of voters will behave, not assign numbers that reflect assessments of film quality.

After all, every film critic would agree that what wins Best Picture is not always the year's best movie. Find me the film historian who believes that Best Picture winner *How Green Was My Valley* (1941) is better than its competitor, *Citizen Kane* (1941).

In the heat of Oscar season, there are campaigns, reviews, gossip, and voting deadlines to contend with. There's no time to let the films sink in. There's no time to let the wisdom of history rank the movies. So the voters make their best judgments, and that's all they can do.

With the benefit of hindsight, film experts tend to converge around opinions that often don't agree with the Academy. In 1958, the Academy chose to honor *Gigi*, but didn't even nominate *Vertigo* or *Touch of Evil* for Best Picture. Both are now considered to be among Hollywood's finest. This doesn't mean that these latter-day experts are more "correct" than the contemporary voters – after all, who's to say what's correct? – but at least they had the opportunity to see which films stood the test of time.

I will now attempt to do the very task commenters have warned me not to do: use math to determine what movie *should* have won Best Picture each year of the Oscars. To do this, I will take an average across prominent critic and viewer polls that enjoy the benefit of hindsight, and then I will compare the results to each year's Best Picture selection.

There have been a number of attempts to rank cinema's best output, and I will choose five of the most prominent to work with. *Sight & Sound* has been putting out a critic poll every decade since 1952, with the latest in 2012. The American Film Institute (AFI) conducted a similar poll of Hollywood in 1998 and 2007. I'll use the most recent standings from *Sight & Sound* and the AFI. *The Hollywood Reporter* released its own top 100 poll of industry insiders in 2014. IMDb assigns every movie an average user rating. The Library of Congress' National Film Preservation Board, consisting of filmmakers and academics, has selected 25 films for preservation every year since 1989.

The trouble is, these five lists are not on the same scale. Three of them (*Sight & Sound*, the American Film Institute, and *The Hollywood Reporter*) are rankings of films, averaged across many voters. IMDb publishes a rating of every movie on a 1-to-10 system. The Library of

Congress just tells us the year it preserved the film, without putting those films in order of preference. My solution is to put everything on an IMDb scale.

Across IMDb's entire dataset, there are 8,099 films with at least 10,000 votes each. These films have a mean rating of 7.04 and a *standard deviation* (a common statistical term for how spread out a group of numbers is) of 0.63. Using this information, we can assign a hypothetical IMDb score to the highest movie on a list, the second-highest movie on a list, and so on. This is how we'll handle *Sight & Sound*, the AFI, and *The Hollywood Reporter*. As an example, the AFI ranks *Citizen Kane* (1941) as the #1 movie of all time, which in my model translates into 9.36 points.[1]

For the Library of Congress, we'll rank films by how many years it took each movie to be preserved after it was first eligible. For example, *The Best Years of Our Lives* (1946) was one of the inaugural 25 films selected in 1989, so it was inducted in its first possible year of eligibility. Films are not eligible for the Library's collection until ten years after release. While both *Blade Runner* (1982) and *The Godfather: Part II* (1974) were inducted in 1993, this was *Blade Runner*'s second year of eligibility (following that ten-year waiting period), but *The Godfather: Part II*'s fifth, as *The Godfather: Part II* had been eligible since the opening set of 1989. Hence, *Blade Runner* ranks higher on this list. This gives us a ranking of all movies in the Library's archive, which I also converted to the IMDb scale.

A couple of notes before getting into the results: First, I am taking a simple average of these five lists, which means all five get equal

---

[1] In a normal dataset of mean 7.04 and standard deviation 0.63, statistics tells us that the threshold which we'd expect one film per 8,099 to reach is 9.36 (look up the *normal distribution* if you'd like to learn more). So, *Citizen Kane* gets 9.36 points in my model from its AFI ranking.

weight. No weighting is inherently correct. Some people might prefer to assign more weight to moviegoers, in which case they'd favor weighting IMDb more heavily. Others might choose to give more weight to critics, in which case they'd value the critic polls more highly. Both are equally valid, so I'm compromising by allocating equal weight to all five.

Second, this ranking of course carries with it the biases of the underlying rankings. If you feel the IMDb list is too high on superhero films, you'll think this list shares that problem. If you don't like that the AFI list is primarily focused on American cinema, you'll notice the same issue here. However, none of those problems will be quite as pronounced as in the original lists because each of the underlying lists only accounts for one-fifth of the overall ranking.

Without further ado, let's go back in time to determine with the benefit of hindsight which movie should have won Best Picture each year, according to thousands of reviews and a hefty dose of math. This method also produces a ranking of all movies across years, so when I mention a film that is in the overall top 100, I will include its all-time ranking before the film's title. The full top 100 list appears at the end of this chapter. An asterisk on a should-have-won movie indicates the film was not even nominated for Best Picture.

1928

Winner: *Wings*

Should have won: #34 *Sunrise** (it won an Oscar in a parallel category, "Best Unique and Artistic Production")

Technically, the Academy did honor *Sunrise* as the Best Picture of the year, a title it most certainly deserves for its at-times charming and at-times heartbreaking story of a couple in love. However, the inaugural ceremony awarded two Best Picture honors, and the

Academy retroactively declared the one won by engineering marvel *Wings* the predecessor to the modern Best Picture category. In the same category as *Sunrise*, the Academy also nominated King Vidor's #88 *The Crowd*, which has stood up well over the decades thanks to its sometimes joyful, sometimes tragic, but always insightful look at the anonymity of life in New York.

1929

Winner: *The Broadway Melody*
Should have won: *The Wind**

As far as timing, it couldn't have been worse for *The Wind*, Lillian Gish's final silent film about a woman who is nearly driven insane by an incessant wind. It was the dawn of the sound era, and the Academy was eager to honor music-filled *Broadway Melody*. But in retrospect, many critics call *The Wind* one of the best films of its decade.

1930 – Correct!

Winner: *All Quiet on the Western Front*

For the first time in three tries, the Academy got it "right," at least by modern standards. Based on the bestselling World War I novel, *All Quiet on the Western Front* is a moving look at soldiers during a fight that was still very fresh in audiences' minds in 1930.

1931

Winner: *Cimarron*
Should have won: #35 *City Lights**

According to my metric, this is the biggest mistake in Oscar history. After getting its first "correct" answer with *All Quiet on the Western Front* (1930), the Academy reverted to its old ways. As in 1929, the relatively dull sound film *Cimarron* beat out what, in my opinion, is the greatest silent film of all time, Charlie Chaplin's *City Lights*. Though to be clear, the list of should-have-won films in this chapter is based purely upon the math described above.

1932

Winner: *Grand Hotel*
Should have won: *Freaks**

Freaks had the courage to depict humans with deformities as real people with real feelings, while making able-bodied circus performers the villains. Apparently, this concept was so horrifying to audiences at the time that MGM pulled it from distribution and the United Kingdom banned it for three decades. The Academy opted for the comparably safe choice, *Grand Hotel*. But many years later, *Freaks* became something of a cult classic, finally finding its fans in an era more welcoming to films outside the mainstream.

1933

Winner: *Cavalcade*
Should have won: #74 *Duck Soup**

Cavalcade, which tells the story of an early nineteenth-century British family, feels much weightier than the hilarious Marx Brothers comedy *Duck Soup*. But weightier doesn't always mean better, and in hindsight, nearly all critics have come to agree that the Academy got it wrong in 1933. *Duck Soup* is virtually tied with the innovative beauty-and-beast story #78 *King Kong* and the delightful thieves-in-love tale #81 *Trouble in Paradise*.

1934 – Correct!
Winner: #90 *It Happened One Night*

A year after overlooking the rollicking *Duck Soup*, the Academy opted for the lighthearted choice in 1934. Frank Capra's *It Happened One Night* tells the story of a journalist on the road who falls in love with his article's subject. This charming picture captured the hearts of voters, becoming the first of three movies to ever win Best Picture, Director, Actor, Actress, and Adapted Screenplay. (*One Flew Over the Cuckoo's Nest* in 1975 and *The Silence of the Lambs* in 1991 are the others.)

1935
Winner: *Mutiny on the Bounty*
Should have won: *A Night at the Opera**

Once again, the Marx Brothers were snubbed for their work in *A Night at the Opera*. It's not as if the Academy was averse to all comedy – between 1933 and 1935, it correctly honored *It Happened One Night* – but perhaps the Marx Brothers were just a bit too silly to get the Best Picture recognition they deserved. Instead, the Academy honored the Clark Gable vehicle *Mutiny on the Bounty*, the most recent film to win Best Picture and nothing else.

1936
Winner: *The Great Ziegfeld*
Should have won: #42 *Modern Times**

With an epic musical biopic in *The Great Ziegfeld* and Charlie Chaplin's silent classic *Modern Times*, you can't go wrong, but *Modern Times* is, in my opinion (and that of myriad critics and viewers), the better film. However, by this point, Chaplin's insistence on staying silent in an industry that had long since started talking meant he never had a chance at a nomination, let alone a win.

1937

Winner: *The Life of Emile Zola*
Should have won: *Snow White and the Seven Dwarfs**

    The Life of Emile Zola contains one of the most moving pleas for justice ever uttered in cinema, as a French journalist defends a falsely accused Jewish man. But in terms of significance, that's nothing compared to *Snow White*, which catapulted an entirely new genre into the limelight – the animated feature.

1938

Winner: *You Can't Take It with You*
Should have won: #71 *Bringing Up Baby**

    What a year for laughs! Take your pick: *Bringing Up Baby*, my favorite screwball comedy ever; *You Can't Take It with You*, the Academy's pick and another Frank Capra gem; *The Adventures of Robin Hood*, a swashbuckling adventure; *Love Finds Andy Hardy*, an adorable Mickey Rooney/Judy Garland romance; and plenty more.

1939

Winner: #17 *Gone with the Wind*
Should have won: #12 *The Wizard of Oz*

    1939 is commonly cited as the greatest year in film history. You know it's a strong year when *Gone with the Wind* is somehow the "wrong" Best Picture choice, though the films set in Oz and Tara are nearly tied. This is the only year with multiple films in the top 20. A third film from that legendary year just barely cracks the top 100, #98 *Mr. Smith Goes to Washington*, which is my personal favorite film of all time.

1940

Winner: *Rebecca*

Should have won: #50 *The Grapes of Wrath*

This is ironic. Modern-day film lists would have awarded Best Picture to Alfred Hitchcock three times: *Rear Window* (1954), *Vertigo* (1958), and *Psycho* (1960). None of those films were even nominated. Instead, Hitchcock's lone win came for *Rebecca*, a film that present-day critics would have overlooked in place of John Ford's Steinbeck adaptation, *The Grapes of Wrath*, or possibly Charlie Chaplin's first sound film, #86 *The Great Dictator*.

1941

Winner: *How Green Was My Valley*

Should have won: #2 *Citizen Kane*

Really, Academy? I know it's a hard job to recognize timelessness at the time, but this one is tough to justify. At that point, *Citizen Kane* was the best movie ever made according to the metric used in this chapter, only surpassed to this day by *The Godfather* (1972). Orson Welles practically invented modern filmmaking in 1941, yet the troubled coal mining town came out on top. Even if the Academy wanted to overlook the remarkably clever rise-and-fall biopic, two other top 100 choices were in the running that year: film noir essential #62 *The Maltese Falcon* and ode to the value of comedy #72 *Sullivan's Travels*.

1942

Winner: *Mrs. Miniver*

Should have won: #96 *The Magnificent Ambersons*

By this point, the Oscars were just being cruel to Mr. Welles, robbing him of not one but two Best Pictures in back-to-back years. Welles' *The Magnificent Ambersons* is a harrowing riches-to-rags story that perhaps could have been even more powerful had RKO not chopped off nearly an hour of the film over Welles' objections. Not only is that edited footage lost, but the film itself lost to *Mrs. Miniver*, a spectacular movie in its own right about the tenacity of the British during World War II, made at a time when the outcome of that war was still very much in doubt.

1943 – Correct!

Winner: #4 *Casablanca*

"Here's looking at you, kid." "Louis, I think this is the beginning of a beautiful friendship." "Play it, Sam. Play 'As Time Goes By.'" "Round up the usual suspects." "We'll always have Paris." "Of all the gin joints in all the towns in all the world, she walks into mine." And so on, and so on. If it's possible to write a perfect script, the Epstein brothers and Howard Koch achieved just that in 1942, and credit the Academy for recognizing them accordingly the following year. All six of those quotes were included in the American Film Institute's 100 greatest movie quotes; no other film exceeds three.

1944

Winner: *Going My Way*

Should have won: #61 *Double Indemnity*

Double Indemnity, about a woman conspiring to murder her husband to collect on his life insurance, is one of the keystone movies of the film noir genre. The Academy definitely has a pattern of overlooking this genre from time to time, but *Going My Way* is itself a masterpiece, albeit less celebrated or remembered. Bing Crosby, playing a heart-of-gold priest, comes seemingly out of nowhere to transform a parish and all of its inhabitants for the better. And of course, he croons throughout it all.

1945

Winner: *The Lost Weekend*
Should have won: *Mildred Pierce*

These movies are quite close to each other for the top spot of the year, so the Academy didn't necessarily get this wrong. *Mildred Pierce* is an excellent movie about a mother covering for her daughter's murder charge, while *The Lost Weekend* is a powerful portrayal of the horrors of alcoholism. Joan Crawford delivers a knockout performance in the former, while Ray Milland does some of the finest acting of the decade in the latter. Both won Oscars for their respective roles.

1946

Winner: #57 *The Best Years of Our Lives*
Should have won: #19 *It's a Wonderful Life*

What a top two. Modern-day audiences, including me, prefer *It's a Wonderful Life* (incidentally, another Frank Capra/Jimmy Stewart inspirational story, like *Mr. Smith Goes to Washington*). But you can't blame the Academy for picking *The Best Years of Our Lives*, the saga of three soldiers returning to life on the home front that's almost as moving now as it was to audiences watching in theaters shortly after the war ended.

1947

Winner: *Gentleman's Agreement*
Should have won: #97 *Out of the Past**

Out of the Past hasn't enjoyed the long-lasting popularity of other film noir classics (*The Maltese Falcon* and *Double Indemnity* come to mind), but among historians, it's widely considered to be one of the best examples of the genre. It does require sifting through a confusing plotline and one too many cigarettes, but in exchange the viewer is rewarded with nonstop tension and shadows. The Academy overlooked it, going with the powerful story in *Gentleman's Agreement* of a reporter who pretends to be Jewish in order to uncover anti-Semitism in his community.

1948

Winner: *Hamlet*
Should have won: #63 *The Treasure of the Sierra Madre*

Hamlet, of course, is Shakespeare's famous play about the prince of Denmark. Laurence Olivier plays the title role in his typically stoic manner. It's as good as any other performance of the iconic show. *The Treasure of the Sierra Madre* is far more original, a parable about greed disguised as a Western, and in my view is the second-greatest Western ever, trailing only *High Noon* (1952).

1949

Winner: *All the King's Men*
Should have won: *White Heat*

Made it, Ma, top of the ranking. James Cagney plays a double-crossed gangster in the explosive cops-and-robbers classic *White Heat*, and films historians have named it the best film of 1949. Though for relevance in the modern world, the Academy's choice of *All the*

King's Men makes a strong case. This fictionalized version of the life of Louisiana Governor Huey Long serves as an important warning about corruption in government both then and now.

1950
Winner: #28 *All About Eve*
Should have won: #10 *Sunset Boulevard*

There is only one other year on the entire list in which both the winner and a different film that should have won reside in the top 28, where *All About Eve* sits. You guessed it: 1939. Here we have a true heavyweight battle between the story of an ambitious actress at the beginning of her career in *All About Eve* and the tale of a spotlight-hungry actress well past her prime in *Sunset Boulevard*. Experts and audiences have come around to *Sunset Boulevard*, which is likely the correct call, but there isn't a bad choice here.

1951
Winner: *An American in Paris*
Should have won: *A Place in the Sun*

This was an upset even at the time, and it's still true today. The Oscars opted for Gene Kelly and Leslie Caron dancing through *An American in Paris* over the expected winner, compelling upper-class drama *A Place in the Sun*. Also overlooked that year were *The African Queen*, which wasn't even nominated, and *A Streetcar Named Desire*.

1952
Winner: *The Greatest Show on Earth*
Should have won: #7 *Singin' in the Rain**

The Greatest Show on Earth does not appear on any of the film rankings I used to create this model, while *Singin' in the Rain* sits high up on all of them, even clocking in at #5 according to AFI. If you had to pick one musical emblematic of the genre's rise in the early 1950s, you'd be hard-pressed to find a better example. But personally, I feel *The Greatest Show on Earth* deserves at least a little love from the eyes of history, at the very least for its innovative use of actual circus members and sets. The Academy also could have made an excellent call with #95 *High Noon*, a brilliant real-time analysis of how society reacts to an emergency.

1953

Winner: *From Here to Eternity*
Should have won: *Shane*

No 1953 film reaches the top 100, so the Academy was left to choose between two movies that each have one iconic scene. *From Here to Eternity*, about lovelorn soldiers in Hawaii, is best remembered for the kiss shared by Burt Lancaster and Deborah Kerr as the waves rush over them. *Shane* is strongly associated with its poignant final moments as a young boy cries out over the Western frontier for Shane to come back. At that Oscar ceremony, American soldiers won the battle.

1954

Winner: #31 *On the Waterfront*
Should have won: #27 *Rear Window**

It's really hard to call this one a mistake. While Alfred Hitchcock's *Rear Window* is technically ahead, it's virtually tied with the Academy's choice, *On the Waterfront*. The latter is a moving tale of a corrupt dockworkers union, featuring Oscar-winning direction by Elia Kazan, Oscar-winning acting by Marlon Brando and Eva Marie Saint, and a score by renowned composer Leonard Bernstein.

1955

Winner: *Marty*
Should have won: #87 *The Night of the Hunter**

Though I did not include the *Cahiers du Cinema* poll among the five I averaged to create these rankings, it's worth noting that in 2008, the French film magazine ranked *The Night of the Hunter*, about a serial killer trying to steal bank robbery loot from another thief's widow and her children, at number two on its list after *Citizen Kane*. Overall, my compilation does not produce quite so generous a result, but the film does crack the top 100, unlike the actual 1955 winner *Marty*, the simple home-movie-esque love story.

1956

Winner: *Around the World in 80 Days*
Should have won: #32 *The Searchers**

The Searchers is the quintessential Western of bygone years: It's directed by John Ford, stars John Wayne, and features the grossly stereotypical depiction of enmity between purportedly good white men and evil Native Americans. At the Oscars, it was shut out entirely in favor of an "It's a Small World"-style tour of world culture, *Around the World in 80 Days*. And though it wasn't nominated by the America-centered Academy, #73 *Seven Samurai* was also eligible for the Oscars in 1956 and is now considered a landmark of world cinema.

1957

Winner: #68 *The Bridge on the River Kwai*
Should have won: #45 *12 Angry Men*

Both of these films are beloved by modern critics, and I can almost imagine some Academy members needing to flip a coin between the pair. *The Bridge on the River Kwai*, dealing with British prisoners of war and the madness of fighting, is as epic as epic can be. In *12 Angry Men*, a film adaptation of one of the most captivating courtroom plays in history, a lone juror holds out for justice. I strongly recommend watching both to decide for yourself. Also in contention for that year's number one film was #76 *Paths of Glory*, Stanley Kubrick's damning portrayal of military justice.

1958
Winner: *Gigi*
Should have won: #3 *Vertigo**

Vertigo seemingly did the impossible when it dethroned *Citizen Kane* in the latest *Sight & Sound* poll, so it's no surprise that this model likes it so much. I'm not going to disagree – in my view, the Jimmy Stewart-led thriller is Alfred Hitchcock's crowning achievement, and Bernard Herrmann's score is as stirring as they come. The Oscars also could have gone with #85 *Touch of Evil*, Orson Welles' winding film noir best remembered for its explosive tracking-shot opening. *Gigi*, on the other hand, is a disturbing musical about a young French girl being trained as a mistress.

1959
Winner: *Ben-Hur*
Should have won: #13 *Some Like It Hot**

In *Some Like It Hot*, Jack Lemmon and Tony Curtis accidentally witness a mob hit and are forced to escape by dressing up as women. For those who find Billy Wilder's dialogue and plot amusing, *Some Like It Hot* ranks among their favorite films. Personally, I have a tough time blaming the Academy for going with Roman slave epic *Ben-Hur*, even if the math says it got this one very wrong. Aside from *Ben-Hur*, the Academy had another dramatic option, #30 *North by Northwest*, Alfred Hitchcock's heart-pounding all-American chase with an iconic Mount Rushmore finale.

1960
Winner: #64 *The Apartment*
Should have won: #15 *Psycho**

I must admit, I've always wondered if this was redress for Billy Wilder after the Academy failed to nominate *Some Like It Hot* a year before. And the eyes of retrospection would have looked kindly upon that decision if it hadn't been for another Hitchcock thriller, *Psycho*. It's the movie that made America afraid to take a shower for weeks and features some of the most terrifying horror scenes ever put on film.

1961 – Correct!
Winner: #83 *West Side Story*

When I attended Harvard, the school offered a popular class on the history of Western music. The only musical on the syllabus? *West Side Story*, which was immensely well-regarded on stage and just as successful on screen.

1962 – Correct!
Winner: #16 *Lawrence of Arabia*

This was quite a year for one-person-against-the-world films. The Academy correctly picked *Lawrence of Arabia*, the World War I adventure that practically defines the word "epic." But that year also saw Atticus Finch take on a racist town in #38 *To Kill a Mockingbird*, based on what is (according to me, in a purely subjective sense) the greatest novel ever written.

1963
Winner: *Tom Jones*
Should have won: *8½**

This year offered a lot of good options. The math says the right call was Federico Fellini's avant-garde semi-autobiographical *8½*. Personally, I would add *Lilies of the Field, Charade, It's a Mad, Mad, Mad, Mad World, Bye Bye Birdie,* or any number of other 1963 choices to the mix. I might not include the Academy's choice, *Tom Jones,* an odd eighteenth-century comedy that was considered very amusing at the time.

1964
Winner: *My Fair Lady*
Should have won: #18 *Dr. Strangelove*

Dr. Strangelove is a strange comedy that makes fun of the perfectly natural Cold War fear of nuclear war. It's a love-it-or-hate-it kind of film, and enough voters love it that it's often found near the top of lists of the best movies of all time. It was up against a pair of beloved musicals, *My Fair Lady* and *Mary Poppins*, each with magnificent lead actresses in Audrey Hepburn and Julie Andrews, respectively. *My Fair Lady* was Warner Brothers' most expensive movie to date and it paid off big, both at the box office and at the Academy Awards.

1965 – Correct!
Winner: #53 *The Sound of Music*

Thanks in large part to this being my mom's favorite film, *The Sound of Music* is still the movie I have seen more than any other. And if you have to repeat one movie over a dozen times, it's hard to pick a better choice. From an incredible Rodgers and Hammerstein score to a powerful true backstory, this may be the greatest movie musical ever made, and a worthy Best Picture then and now.

1966

Winner: *A Man for All Seasons*
Should have won: *Who's Afraid of Virginia Woolf?*

After *Citizen Kane* (1941), *The Magnificent Ambersons* (1942), *The Third Man* (1949), and *Touch of Evil* (1958), a movie featuring Orson Welles finally wins Best Picture. Of course, he had much less to do with *A Man for All Seasons*, playing the supporting part of Cardinal Wolsey in the adaptation of Robert Bolt's play about Sir Thomas More. But *Who's Afraid of Virginia Woolf?* triumphs on the should-have-won list with a shouting-filled drama about living with illusions.

1967

Winner: *In the Heat of the Night*
Should have won: #44 *The Graduate*

At the dawn of the counterculture movement, the top films of 1967 laid out a fairly depressing view of America. There was a strong contender in criminals-are-just-like-us #54 *Bonnie and Clyde*. There was the movie that should have won according to modern polling, suburban-life-is-meaningless *The Graduate*. And there was the movie that did win, exposé-of-southern-racism *In the Heat of the Night*. Sometimes, dark times lead to superb moviemaking.

1968

Winner: *Oliver!*
Should have won: #6 *2001: A Space Odyssey**

Surely one of the most unique films ever made, *2001: A Space Odyssey* is a series of largely unrelated segments, most memorably the one in which a space shuttle's computer system turns against its human inhabitants. Love it or hate it, there's no denying its originality.

On the other hand, *Oliver!* was the fifth musical over the previous decade to win Best Picture.

1969

Winner: *Midnight Cowboy*
Should have won: #60 *Butch Cassidy and the Sundance Kid*

One of these two movies famously concludes with a violent outburst of gunfire. And yet, it's the *other* movie that still to this day is the only X-rated Best Picture winner. Granted, when *Midnight Cowboy* received that X-rating, it didn't quite have the same connotation that it does now. (Today, *Midnight Cowboy* would be rated R, as are many other Best Pictures.) Even so, this year solidified that the late-1960s trend of dark realism in cinema was not a fad, but here to stay. Further evidence: This year also saw the release of the dark Western #92 *The Wild Bunch*.

1970

Winner: *Patton*
Should have won: *Five Easy Pieces*

Both of these films earned nominations, and neither has entered the overall top 100. Modern critics prefer *Five Easy Pieces*, about a blue-collar man with a white-collar upbringing. Voters at the time went with *Patton*, a biopic of the World War II general.

1971

Winner: *The French Connection*
Should have won: #58 *A Clockwork Orange*

At the very least, I must give credit to the Academy for even nominating *A Clockwork Orange*. As a haunting, brilliant look into society's attempts to reform a teenage criminal, it's not exactly standard Oscar fare. The members had the guts to nominate it, but no more: In the end, the award went to a much more conventional film, crime drama *The French Connection*, most notable for its incredible car chase scene through New York.

1972 – Correct!
Winner: #1 *The Godfather*

In hindsight, this might seem incredibly obvious. But at the time, there was some intrigue heading into Oscar night. *The Godfather* and *Cabaret* tied for the most nominations with ten, and prior to the announcement of the final award, *Cabaret* had won eight awards to *The Godfather*'s two. What's more, *Cabaret* had just won Best Director for Bob Fosse, who beat out Francis Ford Coppola of *The Godfather*. Nevertheless, for the most important category, the Academy got it right.

1973
Winner: *The Sting*
Should have won: *Badlands**

The Academy overlooked one gritty crime drama (Terrence Malick's *Badlands*) in favor of another, the complicated Depression-era movie *The Sting*. Not that the voters knew it at the time, but this would be their last complete miss for a while. Over the next eight years, the best film of each year would at least receive a nomination, and three of them won – the most accurate era in Academy history according to the modern polls.

1974 – Correct!
Winner: #5 *The Godfather: Part II*

The Academy chose correctly in a very difficult year to do so. One option was #23 *Chinatown*, another Best Picture nominee, about crime and corruption in 1930s L.A. Another was #70 *A Woman Under the Influence*, about a housewife undergoing psychiatric treatment. But best of all, there was *The Godfather: Part II*, the part-prequel/part-sequel to the iconic 1972 Best Picture winner that is every bit its worthy successor.

1975 – Correct!
Winner: #22 *One Flew Over the Cuckoo's Nest*

I don't know if you've noticed yet, but we're in the midst of an amazing stretch. Even though only eight years have at least three films on the top 100, three of those years occur consecutively: 1974, 1975, 1976. With a new classic coming out every few months, you might think it would be harder for the voters to make the right call, but they did it again in 1975, according to later polls, going with the harrowing psychiatric ward over the iconic shark-attack blockbuster (#46 *Jaws*) and the bicentennial country music tribute (#82 *Nashville*).

1976
Winner: #94 *Rocky*
Should have won: #24 *Taxi Driver*

Ironically, the storyline of *Rocky* is all about an underdog's failure to pull off an improbable victory. Sometimes, real life is even more triumphant than art, as the Philadelphia hero won the title bout at the Oscars over Martin Scorsese's iconic *Taxi Driver* as well as #89 *All the President's Men*.

1977
Winner: #29 *Annie Hall*
Should have won: #8 *Star Wars*

Annie Hall is classic Woody Allen and has its funny moments. But let's call a spade a spade: *Annie Hall* did not change the entire landscape of American cinema. *Star Wars* did. Every action film since 1977 owes at least some debt to George Lucas' universe, making it easy to see why this aggregated ranking of history's preeminent films places it in the top ten. It's such an important film that *Close Encounters of the Third Kind*, also released in 1977, had the misfortune to not even be the number one supernatural film of its year, despite being one of the shining examples of the genre.

1978 – Correct!
Winner: #46 *The Deer Hunter*

The Deer Hunter is best remembered for its brutal Russian roulette scenes in which American prisoners of war are forced to play the deadly game. It's a chilling metaphor for war, but also a questionable plotline given the absence of records of this particular type of abuse during the Vietnam War. Historical accuracy aside, it's a well-made film and a worthy Best Picture.

1979
Winner: *Kramer vs. Kramer*
Should have won: #14 *Apocalypse Now*

Kramer vs. Kramer stars Meryl Streep and Dustin Hoffman, but it's eight-year-old Justin Henry who steals the show. All three were nominated for their performances, with the two adults winning. It's a relatively ordinary story about a custody battle and couldn't be more different from *Apocalypse Now*, a surreal take on the Vietnam War that's really a nightmare masquerading as a movie. Speaking of nightmares, 1979 also saw the release of Ridley Scott's terrifying space drama, #49 *Alien*.

1980

Winner: *Ordinary People*
Should have won: #9 *Raging Bull*

Rather than award Best Picture to a second boxing film in five years, the Academy opted for its second depressing family drama in two years. If it's any consolation to Martin Scorsese, history has strongly vetoed that choice. By 1990, the Library of Congress had admitted *Raging Bull* after the minimum ten-year waiting period, and the compliments haven't died down since. In a violent and disturbing film for the ages, Robert De Niro portrays boxer Jake LaMotta, fighting himself as much as his opponents in the ring. Also released in 1980: #51 *The Empire Strikes Back* and #67 *The Shining*, neither of which were nominated for Best Picture.

1981

Winner: *Chariots of Fire*
Should have won: #39 *Raiders of the Lost Ark*

At first glance, this is a battle between two of the most memorable opening scenes in cinema: British runners on a beach with an incredible score behind them versus Harrison Ford stealing an idol from a temple while the temple fights back. But following the opening scenes, *Chariots of Fire* slows to a jog, while *Raiders of the Lost Ark* keeps up the pace for a thrilling and timeless adventure.

1982
Winner: *Gandhi*
Should have won: #26 *Blade Runner**

The Academy had two terrific science fiction choices that year, as *Blade Runner* and #33 *E.T. the Extra-Terrestrial* are nearly tied for the top spot. Instead, the voters opted for neither, not even nominating the head-spinning *Blade Runner*. The ultimate choice was *Gandhi*, the epic story of one of the most important human beings ever to walk the earth.

1983
Winner: *Terms of Endearment*
Should have won: *The Right Stuff*

The math says that the best movie of 1983 was *The Right Stuff*, a documentary-style drama about Chuck Yeager breaking the sound barrier and the Mercury Seven lifting off into space. Though the Academy did nominate *The Right Stuff* and award it four Oscars, voters handed Best Picture to tearjerker *Terms of Endearment*, which doesn't even make the top ten movies of the year by this model.

1984 – Correct!
Winner: *Amadeus*

If you're ever felt mediocre or eclipsed by the presence of true greatness nearby, have I got a movie for you. *Amadeus*, on a surface level, is about Mozart, but it's really about how his contemporary Antonio Salieri had the misfortunate of living in Mozart's shadow.

1985

Winner: *Out of Africa*

Should have won: #48 *Back to the Future**

To be fair, *Back to the Future* is not normal Best Picture fare, so its omission is hardly shocking. Same goes for *The Breakfast Club*, also released this year. Meryl Streep and Robert Redford starred in *Out of Africa*, a much more run-of-the-mill Oscar film with its upper-class romance and methodical pacing. But *Back to the Future* is one of the most beloved stories ever told, as Marty McFly and Doc Brown try desperately to return to 1985. It was perhaps the most ingenious movie plot written to that point, only to be topped by the even more brilliant *Back to the Future Part II* (1989).

1986

Winner: *Platoon*

Should have won: *Ferris Bueller's Day Off**

To be fair to the Academy, this would have been hard to see coming. Teen comedies were a dime a dozen in those days, and it would have been difficult to predict that *Ferris Bueller* would be the one to stand the test of the time. Even the box office, normally a reliable indicator of the popularity of teen comedies, led the Academy astray: The ode-to-truancy *Ferris Bueller* ranked tenth that year, but Vietnam War film *Platoon* came in third.

1987

Winner: *The Last Emperor*

Should have won: *The Princess Bride**

The Last Emperor is an epic tale of the noble birth and ultimate downfall of China's final imperial ruler before the dawn of Communism. It's one of only seven films to win at least nine Oscars, but the one label it couldn't attain was that of cult classic. That status is what propelled the swashbuckling *The Princess Bride*, initially only a modest success, into long-term stardom.

1988

Winner: *Rain Man*
Should have won: *Die Hard**

Rain Man tells the emotional story of two long-lost brothers and is a worthy Best Picture winner. But *Die Hard*, well, it's just fun. It's the very reason we go to the movies: Bruce Willis against the world, exhilarating action scenes, and an unprintable catchphrase for the ages.

1989

Winner: *Driving Miss Daisy*
Should have won: #66 *Do the Right Thing**

Two films dealing with racism: one with subtlety, one with a shout. The Academy opted for Jessica Tandy's portrayal of an elderly woman who very slowly accepts the presence of a black man in her life in *Driving Miss Daisy*, while hindsight has been kinder to Spike Lee's explosive Brooklyn film *Do the Right Thing*.

1990

Winner: *Dances with Wolves*
Should have won: #11 *Goodfellas*

This shouldn't have been so close. On the one side is a mafia film that almost reached the top ten in this model. On the other side is *Dances with Wolves*, which – how should I say this kindly? – isn't exactly the most gripping movie ever made. Impressively, this would be the Academy's last mistake until four years later – the longest streak of perfection in Oscar history, according to later critics and fans.

1991 – Correct!

Winner: #52 *The Silence of the Lambs*

Horror films don't often get nominated for Best Picture, let alone win. Then again, horror films aren't often (if ever) *Silence of the Lambs*-level quality. Even if you haven't seen this movie in many years, it's possible that the lambs still haven't stopped screaming. And it really had to be that powerful to fend off what, in my humble opinion, is the greatest animated film of all time, *Beauty and the Beast*.

1992 – Correct!

Winner: #79 *Unforgiven*

About a million Hollywood Westerns build their plots around an invincible hero. Far fewer put a decidedly second-rate hero in the saddle. That's *Unforgiven*, the story of an old-fashioned killer who has clearly lost a step but takes one final shot at doing justice. It's just the third Western to win Best Picture, following *Cimarron* (1931) and *Dances with Wolves* (1990).

1993 – Correct!

Winner: #21 *Schindler's List*

No work of art, film or otherwise, could ever possibly capture the full horror of the Holocaust. But the work of art that comes closest may be Steven Spielberg's *Schindler's List*. For those with 1993 nostalgia who are in the mood for something decidedly lighter, I would suggest the brilliantly repetitive comedy #56 *Groundhog Day*.

1994

Winner: #37 *Forrest Gump*

Should have won: #20 *Pulp Fiction*

Here's what's impressive about 1994: Not only is *Forrest Gump*, arguably cinema's most compelling vision of twentieth-century America, not the rightful winner, but it's not even in second. Because in between Tom Hanks' famous role and Quentin Tarantino's brutal *Pulp Fiction* sits #25 *The Shawshank Redemption*, Hollywood's finest portrayal of prison and escape. Add in Disney's classic #99 *The Lion King*, and 1994 becomes the only year with four films in the top 100.

1995

Winner: *Braveheart*

Should have won: #36 *Toy Story**

Historical accuracy be darned, *Braveheart* is still a breathtaking version of dramatized Scottish history. But *Toy Story*, like *Snow White* before it, basically invented a whole new genre: the computer-animated feature film. This genre has since found immense box office riches despite failing to win Best Picture thus far. Also award-worthy from 1995: #93 *The Usual Suspects* – if you haven't seen it, please, please put down this book and watch it immediately before someone ruins the ending for you.

1996

Winner: *The English Patient*

Should have won: #43 *Fargo*

The English Patient is a quintessential Oscar film in the stereotypical sense: a not-very-straightforward love story set during World War II with sweeping Middle Eastern vistas and lots of star acting performances, many by performers with British accents. The Academy ate it up, delivering 9 wins on 12 nominations. But *Fargo*, silly and violent in a way that only the Coen brothers can muster, has already earned a longer shelf life, just two decades after the dueling films were released.

1997 – Correct!
Winner: #91 *Titanic*

I'm a huge *Titanic* fan and would go so far as to say that from a purely technical perspective (filming, effects, sound, etc.), it deserves a place in the discussion on the best-produced film in Hollywood history. The Academy agreed, giving it a record-tying 11 wins on a record-tying 14 nominations. The overall ranking isn't quite as generous as I am, but #91 isn't too shabby and is good enough to make it the rightful champion in 1997.

1998
Winner: *Shakespeare in Love*
Should have won: #59 *Saving Private Ryan*

This is one of the Academy's most famous errors, recognized as such almost immediately after the envelope was opened. *Saving Private Ryan*'s Normandy battle scene is arguably the finest film depiction of war ever produced: 30 minutes of storming the beach in all of its realistic horror. But I urge you not to criticize the Academy for 1998 unless you've seen both films. *Shakespeare in Love* is basically a series of incredibly witty Shakespeare puns, and if (like me) you enjoy that sort of shtick, you'd agree it's a reasonable Best Picture choice. Speaking of wit, 1998 also saw the release of #84 *The Big Lebowski*, whose standing is hurt by being the only American film on the top 100 to not receive one of the AFI's 400 nominations.

1999
Winner: *American Beauty*
Should have won: #41 *The Matrix**

From *The Sixth Sense* to #77 *Fight Club* to *The Matrix*, 1999 is the ultimate year of plot twists. To its credit, the Academy at least had the good sense to nominate one of these, *The Sixth Sense*, despite its hailing from a genre that is generally overlooked come awards season. However, it didn't have the guts to go all the way, opting for the more conventional *American Beauty*, a commentary on the dullness of suburban life.

2000 – Correct!
Winner: *Gladiator*

Much like *Braveheart* five years before it, *Gladiator* is the epic story of a man hell-bent on avenging his family. Neither could be accused of documentary-level historical accuracy, for what that's worth, but both were excellent films and solid choices for Best Picture.

2001
Winner: *A Beautiful Mind*
Should have won: #75 *Memento**

Memento is so thoroughly ingenious that it requires multiple viewings to piece it all together. Still, that wasn't enough to topple *A Beautiful Mind*, the story of Princeton professor John Nash's struggle with schizophrenia. This year also began the run of dominance by Peter Jackson's famous trilogy, as #80 *The Lord of the Rings: The Fellowship of the Ring* reached the top 100 and just barely missed surpassing *Memento*.

2002

Winner: *Chicago*

Should have won: *The Lord of the Rings: The Two Towers*

For those who love the Middle-earth trilogy that kicked off at least two decades of fantasy action movies, the fact that *Lord of the Rings* rates higher than Oscar winner *Chicago* is welcome news. For others, you can take comfort in the fact that *The Two Towers* is one of only two should-have-won films in this chapter (along with *The Right Stuff*) that didn't make the top 300.

2003 – Correct!

Winner: *The Lord of the Rings: The Return of the King*

That's right, the math would have picked *Lord of the Rings* to win Best Picture in back-to-back years. It is quite possible that some voters supported part three as a way of honoring the entire trilogy. That said, there was no higher-ranking film in 2003, so whether or not this award was intended to acknowledge three years' worth of achievement, it was still the right call.

2004

Winner: *Million Dollar Baby*

Should have won: *Eternal Sunshine of the Spotless Mind**

Eternal Sunshine of the Spotless Mind, at #102, just barely missed beating out the first *Lord of the Rings* film for the title of most recent film in the top 100. It's a clever tale of a man using technology to wipe his memory clean of a previous relationship and the moral ramifications of that process. The Academy failed to nominate it, instead going for the tragic boxing story *Million Dollar Baby*.

2005

Winner: *Crash*

Should have won: *Brokeback Mountain*

From the moment *Crash* was announced, commenters were already clamoring to name it one of the worst Best Picture decisions ever. That might be a stretch – critics prefer *Brokeback Mountain*, a deeply moving story of two men in love, over *Crash*, an exploration of racism in present-day Los Angeles – but the disparity isn't quite as large as, say, *How Green Was My Valley* versus *Citizen Kane*.

2006

Winner: *The Departed*

Should have won: *Pan's Labyrinth**

Guillermo del Toro's Spanish fairytale *Pan's Labyrinth* did well at the Oscars, picking up three wins on six nominations, quite a haul given the Academy's historical bias against foreign cinema. But it failed to achieve that coveted Best Picture nomination, leaving the door open for Martin Scorsese to finally win his long-denied Best Picture trophy for Boston mob film *The Departed*. Del Toro would have his just reward a decade later, winning Best Picture for the Cold War fantasy *The Shape of Water* (2017). It is at least a decade too early to gauge the place of *The Shape of Water* in film history, but I'm optimistic that future viewers will treat it kindly.

• • •

This model does not cover the era since 2007, as the AFI list and the Library of Congress don't include those films yet. It's very difficult to guess without the benefit of hindsight which films will be remembered most fondly. But in order to complete this discussion, there are three years from the past decade that ought to be mentioned:

In 2008, *Slumdog Millionaire* won Best Picture. Though I'm biased by how much I liked that film, it's already starting to become clear that *The Dark Knight* will enjoy a longer shelf life. After all, *The Dark Knight*'s omission from the Best Picture shortlist was a huge part of the reason that the Academy expanded the category to ten nominees the following year.

In 2009, Oscar winner *The Hurt Locker* was a timely Iraq War movie. But now, both the innovative science fiction film *Avatar* and the rare animated Best Picture nominee *Up* might be en route to eventual vindication.

In 2010, to no one's surprise, *The King's Speech* became yet another British drama to win Best Picture. This masterful film benefits from Colin Firth's acting and a stirring classical score. However, there is a reasonable chance that future critics will remember clever sci-fi nominee *Inception* more favorably. Don't rule out *Toy Story 3*, either.

For readers paying unbelievably close attention, you may have noticed that only 95 films ranked in the top 100 are mentioned above. The explanation is that five silent films on the top 100 predate the Academy Awards:

- #40 *The General* (1926): Buster Keaton stars in this hilarious adventure as a train engineer who performs some remarkable stunts.

- #55 *Intolerance* (1916): D.W. Griffith directed this epic interweaving of four stories of injustice across the ages.

- #65 *The Gold Rush* (1925): Charlie Chaplin's beloved Little Tramp character goes prospecting during the Klondike Gold Rush.

- #69 *Sherlock Jr.* (1924): Another Buster Keaton gem. This time, he's a film projectionist (and wannabe detective) who falls asleep on the job and finds himself transported inside the movies he's showing.

- #100 *Greed* (1924): Erich von Stroheim wrote and directed an epic tale of romance, jealousy, and tragedy that vividly demonstrates how a desire for money can lead to moral decay.

A number of the above films currently listed on the top 100 have to come off the list sooner or later. We have a whole 'nother century of films coming up, surely filled with more all-time classics which will displace some of the current entries. Hopefully, the Academy correctly honors as many of those classics as possible. But even if it doesn't, both the right and wrong choices will provide us with the endless pleasure of debating the greatness of movies.

• • •

Here is the aggregated top 100 list in order. Best Picture winners are marked in bold and are followed by the word "Winner." Best Picture nominees are marked in italics and are followed by the word "Nominee."

#1. **The Godfather (1972): Winner**
#2. *Citizen Kane (1941): Nominee*
#3. Vertigo (1958)

#4. **Casablanca (1942): Winner**
#5. **The Godfather: Part II (1974): Winner**
#6. 2001: A Space Odyssey (1968)
#7. Singin' in the Rain (1952)
#8. *Star Wars (1977): Nominee*
#9. *Raging Bull (1980): Nominee*
#10. *Sunset Boulevard (1950): Nominee*
#11. *Goodfellas (1990): Nominee*
#12. *The Wizard of Oz (1939): Nominee*
#13. Some Like It Hot (1959)
#14. *Apocalypse Now (1979): Nominee*
#15. Psycho (1960)
#16. **Lawrence of Arabia (1962): Winner**
#17. **Gone with the Wind (1939): Winner**
#18. *Dr. Strangelove (1964): Nominee*
#19. *It's a Wonderful Life (1946): Nominee*
#20. *Pulp Fiction (1994): Nominee*
#21. **Schindler's List (1993): Winner**
#22. **One Flew Over the Cuckoo's Nest (1975): Winner**
#23. *Chinatown (1974): Nominee*
#24. *Taxi Driver (1976): Nominee*
#25. *The Shawshank Redemption (1994): Nominee*
#26. Blade Runner (1982)
#27. Rear Window (1954)
#28. **All About Eve (1950): Winner**
#29. **Annie Hall (1977): Winner**
#30. North by Northwest (1959)
#31. **On the Waterfront (1954): Winner**
#32. The Searchers (1956)
#33. *E.T. the Extra-Terrestrial (1982): Nominee*

#34. Sunrise (1927): Won parallel category
#35. City Lights (1931)
#36. Toy Story (1995)
#37. Forrest Gump (1994): Winner
#38. To Kill a Mockingbird (1962): Nominee
#39. Raiders of the Lost Ark (1981): Nominee
#40. The General (1926): Pre-Oscars
#41. The Matrix (1999)
#42. Modern Times (1936)
#43. Fargo (1996): Nominee
#44. The Graduate (1967): Nominee
#45. 12 Angry Men (1957): Nominee
#46. The Deer Hunter (1978): Winner
#47. Jaws (1975): Nominee
#48. Back to the Future (1985)
#49. Alien (1979)
#50. The Grapes of Wrath (1940): Nominee
#51. The Empire Strikes Back (1980)
#52. The Silence of the Lambs (1991): Winner
#53. The Sound of Music (1965): Winner
#54. Bonnie and Clyde (1967): Nominee
#55. Intolerance (1916): Pre-Oscars
#56. Groundhog Day (1993)
#57. The Best Years of Our Lives (1946): Winner
#58. A Clockwork Orange (1971): Nominee
#59. Saving Private Ryan (1998): Nominee
#60. Butch Cassidy and the Sundance Kid (1969): Nominee
#61. Double Indemnity (1944): Nominee
#62. The Maltese Falcon (1941): Nominee
#63. The Treasure of the Sierra Madre (1948): Nominee

#64. **The Apartment (1960): Winner**

#65. The Gold Rush (1925): Pre-Oscars

#66. Do the Right Thing (1989)

#67. The Shining (1980)

#68. **The Bridge on the River Kwai (1957): Winner**

#69. Sherlock Jr. (1924): Pre-Oscars

#70. A Woman Under the Influence (1974)

#71. Bringing Up Baby (1938)

#72. Sullivan's Travels (1941)

#73. Seven Samurai (1954)

#74. Duck Soup (1933)

#75. Memento (2000)

#76. Paths of Glory (1957)

#77. Fight Club (1999)

#78. King Kong (1933)

#79. **Unforgiven (1992): Winner**

#80. The Lord of the Rings: The Fellowship of the Ring (2001): Nominee

#81. Trouble in Paradise (1932)

#82. **West Side Story (1961): Winner**

#83. Nashville (1975): Nominee

#84. The Big Lebowski (1998)

#85. Touch of Evil (1958)

#86. The Great Dictator (1940): Nominee

#87. The Night of the Hunter (1955)

#88. The Crowd (1928): Nominated in parallel category

#89. All the President's Men (1976): Nominee

#90. **It Happened One Night (1934): Winner**

#91. **Titanic (1997): Winner**

#92. The Wild Bunch (1969)

#93. The Usual Suspects (1995)

#94. Rocky (1976): Winner

#95. High Noon (1952): Nominee

#96. The Magnificent Ambersons (1942): Nominee

#97. Out of the Past (1947)

#98. Mr. Smith Goes to Washington (1939): Nominee

#99. The Lion King (1994)

#100. Greed (1924): Pre-Oscars

Epilogue. Predicting the Oscars

> KEYES:
>
> Now look, Walter. A guy takes out an accident policy that's worth $100,000 if he's killed on the train. Then, two weeks later, he is killed on the train. And not in a train accident, mind you, but falling off some silly observation car. You know what the mathematical probability of that is? One out of, oh, I don't know how many billions.
>
> –*Double Indemnity* (1944): Nominated for 7 Oscars; won 0

IN 2012, the first year I published Oscar predictions, I received an email in February from a member of the Academy. In a nutshell, he told me I was wrong to project that *The Artist* (2011) had the highest percentage chance of winning Best Picture. He said that among his friends, many of whom were also voting members of the Academy, he was hearing serious backlash against the idea that *The Artist* was such a predictable winner. As such, he divined a groundswell of support for *Hugo* (2011).

I politely thanked him for his comments. Even if I had wanted to adjust my formula to incorporate his feedback, it would have been nearly impossible without breaking my promise to use only numerical data, no opinions. Granted, some of those numbers, like Rotten Tomatoes scores, represent people's opinions, but they're quantifiable. If he told me that, say, 10 out of 12 members of the Academy he spoke with voted for *Hugo*, that's something I could work with. Essentially, that would serve as a poll, albeit a poll with an extremely small and potentially biased sample. But just saying "some friends" isn't exactly music to a mathematician's ears.

So I stuck to my guns. Even though it was my first year of forecasting, I knew *The Artist* was the favorite. How? *The Artist* swept the major Best Picture precursors: the British Academy of Film and Television Arts (BAFTA), the Directors Guild of America (DGA), the Producers Guild of America (PGA), the Golden Globes, and many others. Predictions like that are easy, and it's thanks to such races that nearly every Oscar prognosticator outperforms what we would expect from a random roll of the dice.

The real challenge occurs when the major precursors disagree, and that's where math can excel. I have gathered thousands of data points spanning a wide variety of potential Oscar predictors in every category save for the three short film races. (There isn't enough data to predict those mathematically.) With that historical data, I use statistics to determine which predictors have traditionally aligned closest with the Oscars in each category. That gives me a set of weights for each predictor, and I apply those weights to the current year's data.

To see this in action, here's how my process would have worked for the 2006 Best Picture race:

Film	Director Nom	Editing Nom	BAFTA Win	BAFTA Nom	DGA Win	DGA Nom	PGA Win	PGA Nom	SAG Win	SAG Nom
Babel	1	1	0	1	0	1	0	1	0	1
The Departed	1	1	0	1	1	1	0	1	0	1
Letters from Iwo Jima	1	0	0	0	0	0	0	0	0	0
Little Miss Sunshine	0	0	0	1	0	1	1	1	1	1
The Queen	1	0	1	1	0	1	0	1	0	0

This is a rough snapshot of what the Oscars look like on my computer, though my actual chart contains more columns with additional predictors. The first column lists the name of the nominee. The next two columns, labeled "Director Nom" and "Editing Nom," indicate whether the film was also nominated in other Oscar categories – in this case, Best Director and Best Film Editing. The remaining columns show the film's results at other awards shows prior to the Oscars: those presented by the BAFTAs, the DGAs, the PGAs, and the SAGs. That last one, the Screen Actors Guild, displays the winner of that organization's Best Cast award, the closest category to the Academy's Best Picture honor.

After studying this chart for a while, you can start to guess what the standings will look like. With no Best Film Editing nomination and no nominations from any of these four outside groups, *Letters from Iwo Jima*, Clint Eastwood's film depicting World War II from the Japanese point of view, is a step behind its competitors.

But the standings among the four other contenders are less obvious. *Babel* and *The Departed* are the only two with nominations in each column. *The Departed* also won the Directors Guild award. *Little Miss Sunshine* took home trophies from the Producers Guild and the Screen Actors Guild, but wasn't as warmly welcomed by the Academy, which left it off the lists for Best Director and Best Film Editing. *The Queen* did get a Best Director nomination and a BAFTA win, but not a Film Editing nod. Which of these four has the best Oscar résumé?

Even if you're a true Oscar devotee and you have a hunch as to which film sits in first place, intuition can't tell you precisely how big that movie's lead is. Enter math. Plugging these columns into a common statistical technique[1] trained on data from 1995 (the first year the Screen Actors Guild handed out a Best Cast award) to 2005 (the final year before the one we're predicting), and then applying an adjustment to force all the results to fall between 0% and 100% and add up to 100%, we get the following standings:

1. *The Departed*: 46%

2. *Little Miss Sunshine*: 20%

3. *Babel*: 19%

4. *The Queen*: 14%

5. *Letters from Iwo Jima*: 1%

Turns out it wasn't all that close. The Directors Guild is such a strong predictor of Best Picture that *The Departed* actually had a commanding lead entering Oscar Sunday. This doesn't mean that *The Departed* was assured the victory – 46% is not 100%. It just means that it had the best chance, but there was still a hefty shot that one of the other four movies would pull off an upset. As it turned out, Martin Scorsese's Boston crime drama managed some clever plot twists on-screen, but there was no surprise ending at the Oscars that year: *The Departed* won Best Picture.

This method I've described to predict the Oscars presents some problems, and as such is an oversimplification of my actual method.

---

[1] Linear regression, which is often the first subject covered in an intro-to-statistics course. This is not the precise method I use, for reasons outlined later in this chapter, but the basic concept holds.

Some predictors have small sample sizes, such as new awards shows that just came on the scene a few years ago. Some data is missing, such as movies that don't have Metacritic ratings. Some data is more predictive than others, such as data from more recent years. The final results need to sum to 100%, and some statistical models don't handle that well. Some of the predictors are strongly correlated with each other, a phenomenon that can lead to double counting. Over the years, I have made improvements to my model to address these issues, but the fundamental concept remains the same: use historical Oscar data to determine the relationship between predictors and the Oscars, and then apply that relationship to this year's nominees.

No matter how much effort I put into refining my model, there will always be gaps between predictions and outcomes. At the end of the day, this model predicts percentages and makes no guarantees. Just because a movie with a small chance wins an award, it doesn't necessarily mean the math was incorrect as to that nominee's chances. Sometimes, events with low probabilities occur in the real world. Even in years when everything seems to be going my way – such as 2018, when my model correctly predicted 20 categories out of the 21 I forecasted – all that really tells us is that my model properly identified the favorites, and a high number of favorites won that year.

So why go to all this effort? If no model can perfectly predict the Oscars every year, what's the point?

To me – and hopefully, to you – this work adds excitement to the awards show we all love. Learning exactly how likely each nominee is to win revs audiences up for the big night, filled with triumphs, favorites, longshots, tight races, and sweeps.

On my computer screen and in the chart displayed earlier in this chapter, the 2006 Oscar season just looks like a bunch of 0s and 1s. Mere data points. A dull Excel file. But those 0s are losses, crushing blows to Oscar hopefuls. And those 1s are victories, career-defining honors for talented people in every corner of Hollywood, often individuals who have worked a lifetime for this recognition.

In that sense, the story of these numbers is the story of the Oscars. And what a story it is.

Acknowledgments

> BERTIE:
> However this turns out, I don't know how to thank you for what you've done.
>
> LIONEL:
> Knighthood?
>
> —*The King's Speech* (2010): Nominated for 12 Oscars; won 4 (Best Picture, Best Director, Best Actor, Best Original Screenplay)

Though I work with numbers for a living, it would be very difficult to count the number of people to whom this book and I owe a debt of gratitude. From every friend who has cheered alongside me as Oscar results came in to every follower on Twitter who has spread my articles around the globe, this project of marrying Oscars and math could never have reached such proportions without so many of you.

A few people stand out in particular. First and foremost, my wonderful family – Mom, Dad, Julie, and Emily. The four of you have been my strongest backers in every endeavor of my life, and this one is certainly no exception. Oscar season doesn't truly start for me until I arrive at our annual marathon to watch all of the Best Picture nominees, and Oscar season doesn't end until we've had our post-ceremony telephone debriefing to talk about how that year's predictions went. And at every step in between, the four of you are anxiously following every guild award, every tweet, every article, and every piece of breaking news because you know how much this means to me. Additional thanks go to my loving grandmother MiMi for her constant support.

I'd also like to thank my agent, Andy Ross, as well as the publisher of *Oscarmetrics*, Bear Manor Media, including Ben Ohmart, Darlene Swanson, and Dan Swanson. Needless to say, this book would not exist without you. I am grateful for the fact that you believed in me and my idea.

This project could never have gotten off the ground without the editors who published and promoted my work through the years. I am extremely appreciative of Gregg Kilday, Matthew Belloni, and Erik Hayden at *The Hollywood Reporter*, Stephanie Goodman at *The New York Times*, Zachary Pincus-Roth at *The Washington Post*, Janice Page at *The Boston Globe*, and the Twitter Moments team for helping me share my work with a wide audience.

Finally, my thanks to the online sources that post so much crucial information which enabled me to write this book. IMDb, Rotten Tomatoes, Box Office Mojo, YouTube, the Academy Awards database, and the websites of the other awards shows referenced in these pages proved particularly valuable.

Index

1-9

101 Dalmatians, 122
12 Angry Men, 221-222, 243
12 Years a Slave, 179
2001: A Space Odyssey, 225, 242
5 Fingers, 201
7th Heaven, 204
8 Mile, 99
8½, 224

A

Aardman, 123, 126
About a Boy, 167
Adaptation., 179, 181
Adele, 89
Adventures of Robin Hood, The, 214
Affliction, 16
African Queen, The, 219
After the Thin Man, 166-167
Ajami, 143
Aladdin, 67, 85, 99, 122
Albert Nobbs, 74
Alexander's Ragtime Band, 78
Ali, Mahershala, 24, 28
Alien, 230, 243
All About Eve, 74, 129, 134, 148, 219, 242

All Quiet on the Western Front, 118, 134, 148, 167, 189, 211
All the Brothers Were Valiant, 75
All the King's Men, 28, 148, 179, 218-219
All the Money in the World, 16
All the President's Men, 56, 59, 201, 228, 244
Allen, Woody, 39, 189, 229
Almost Famous, 177
Altman, Robert, 74
Amadeus, 77, 149, 158, 160, 231
Ameche, Don, 25
American Beauty, 59, 150, 237
American Cinema Editors, 47-48, 201
American Dream, 134
American Hustle, 32
American in Paris, An, 77-80, 86, 148, 202, 219
American Sniper, 111-112, 150, 152
American Society of Cinematographers, 47-48
American Splendor, 164
American Tragedy, An, 202
American Werewolf in London, An, 33
Amores Perros, 143
Anchors Aweigh, 86

Anderson, Maxwell, 167
Andrews, Julie, 224
Animal Logic, 127
Anna and the King of Siam, 167
Anna Christie, 74
Anne of the Thousand Days, 134, 167
Annie, 80
Annie Hall, 38-39, 60, 149, 188-189, 229, 242
Anthony Adverse, 24, 133-134
Apartment, The, 149, 223, 244
Apocalypse Now, 229, 242
Apollo 13, 112, 195, 199-200
Arabian Nights, 66
Arcand, Denys, 142
Argo, 2, 187, 195
Arkin, Alan, 19
Around the World in 80 Days, 119, 135, 137, 148, 221
Arquette, Patricia, 25
Arrival, 112, 158
Arrowsmith, 165
Art Directors Guild, 47-48
Artist, The, 57, 247-248
Ashcroft, Peggy, 20, 175
Ashman, Howard, 92
Assault, The, 142
Atonement, 85
Auntie Mame, 27
Avatar, 11, 56, 107, 112, 150, 240
Avengers, The, 12
Awful Truth, The, 167

B

Babel, 249-250
Baby Doll, 166
"Baby, It's Cold Outside," 93
Bacall, Lauren, vii-viii
Back to the Future, 147, 232, 243
Back to the Future Part II, 232
Bad and the Beautiful, The, 28, 70
Bad Girl, 167
Badlands, 227
Balcony, The, 75
Bale, Christian, 27
Barthelmess, Richard, 180
Batman, 56
Beatles, 90
Beatty, Warren, 25, 68, 197-198
Beautiful Mind, A, 150, 181, 237
Beauty and the Beast, 58, 67, 84, 99, 118-119, 122, 234
"Beauty and the Beast," 67, 99
Becket, 74, 162
Bedknobs and Broomsticks, 55
Beery, Wallace, 205
Before Midnight, 167
Before Sunrise, 167
Before Sunset, 167
Beginners, 16
Beineix, Jean-Jacques, 142
Belle Epoque, 142
Bellfort, Tom, 102
Bells Are Ringing, 80
Ben-Hur, 8, 62-63, 70, 102, 149, 222
Benson, Sally, 167
Beresford, Bruce, 200
Bergman, Ingrid, 65

Berkeley, Busby, 98
Bernstein, Leonard, 220
Berry, Halle, ix
Best Years of Our Lives, The, 28, 148, 209, 217, 243
Betty Blue, 142
Big Chill, The, 74
Big Hero 6, 126
Big House, The, 110
Big Lebowski, The, 236, 244
Big Parade, The, 110
Big Short, The, 27
Binoche, Juliette, vii-viii
Birdman, 1, 5, 63-64, 112, 171, 189
Birdman of Alcatraz, 74
Black Swan, The, 66, 70
Blackboard Jungle, 167
Blade Runner, 209, 231, 242
Blade Runner 2049, 112
Bloodbrothers, 167
Blue Sky, 123
Bob & Carol & Ted & Alice, 71
Bolt, Robert, 225
Bon Jovi, Jon, 91
Bonnie and Clyde, 25, 197, 225, 243
Booth, Shirley, 18
Born on the Fourth of July, 173, 200
Bourne Ultimatum, The, 112
Bowling for Columbine, 157-158
Boyes, Christopher, 101-102
Boyhood, 25
Boys Don't Cry, 177
Branagh, Kenneth, 173
Brandauer, Klaus Maria, 180
Brando, Marlon, 220
Brave, 127

Braveheart, 5, 35-36, 104, 112, 150, 195, 199-200, 235, 237
Breakfast Club, The, 232
Bridge of San Luis Rey, The, 56
Bridge of Spies, 27
Bridge on the River Kwai, The, 148, 162, 221, 244
Bringing Up Baby, 214, 244
British Academy of Film and Television Arts, vii, ix, 41-42, 111, 130, 156, 169-175, 190, 199, 201-203, 248-249
Broadbent, Jim, 28
Broadcast News, 151
Broadway Melody, The, 134, 148, 211
Brody, Adrien, 17-18, 181
Brokeback Mountain, 85, 167, 203, 239
Brooklyn, 167
Brooks, Richard, 167
Brother Sun, Sister Moon, 61
Brown, Clarence, 74
Bugsy, 25
Burns, George, 181
Burton, Tim, 123
Butch Cassidy and the Sundance Kid, 134, 226, 243
Bye Bye Birdie, 80, 224

C

Cabaret, 79-80, 82, 164, 195, 227
Cage, Nicholas, 179
Cagney, James, 218
Cahn, Sammy, 88
California Suite, 166

Call Me by Your Name, 16
Call Me Madam, 80
Camelot, 54, 61
Cameron, James, 101
Campanella, Juan Jose, 143
"Can You Feel the Love Tonight," 91
Capra, Frank, xii, 198, 213-214, 217
Captain America: The Winter Soldier, 11
Cardinal, The, 25, 71
Caron, Leslie, 219
Cars, 127
Cartoon Saloon, 123, 126
Casablanca, 148, 188, 216, 242
Cass, Peggy, 27
Cat Ballou, 180
Cat on a Hot Tin Roof, 166-167
Cavalcade, 134-135, 148, 198, 212
Cell, The, 35
Chakiris, George, 19, 181
Champ, The, 205
Chaplin, Charlie, 110, 202, 212-213, 215, 241
Charade, 224
Chariots of Fire, 85, 135, 149, 230
Chastain, Jessica, 179
Chazelle, Damien, 90
Cher, 152
Chicago, 42, 48, 59, 80, 150, 164, 181, 191-192, 195, 238
Chicken Run, 122
Chico & Rita, 126
Children of a Lesser God, 18
"Chim Chim Cher-ee," 89
Chinatown, 227, 242
Cimarron, 56, 134-135, 148, 165, 211, 234

Cinema Audio Society, 45, 47-48, 112-113
Circus, The, 110
Citizen Kane, 188, 205-207, 209, 215, 221-222, 225, 239, 241
City Lights, 110, 211-212, 243
"City of Stars," 90
City Slickers, 28
Cleese, John, 171
Cleopatra (1934), 63
Cleopatra (1963), 63, 70-71, 104, 118, 120
Clockwork Orange, A, 226, 243
Clooney, George, 25
Close Encounters of the Third Kind, 229
Close, Glenn, 74
Coburn, James, 16, 19
Coco, 125
Cocoon, 25
Coen brothers, 69, 236
Colbert, Claudette, xii
Color of Money, The, 167
"Colors of the Wind," 67, 91
Come Back, Little Sheba, 18
Common, 90
Connelly, Jennifer, 181
"Continental, The," 98
Cooper, Chris, 181
Cooper, Gary, 201
Cooper, Jackie, 205
Coppola, Francis Ford, 227
Copti, Scandar, 143
Costume Designers Guild, 41-42, 45, 47-50
Crash, 59, 203, 239
Crawford, Broderick, 179

Crawford, Joan, 217
Creed, 27
Crimson Tide, 112
Critics' Choice Awards, 41-42
Crosby, Bing, 217
Crossfire, 167
Crouching Tiger, Hidden Dragon, 56, 85, 143, 195, 203
Crowd, The, 211, 244
Cruise, Tom, 173
Crunch Bird, The, 115-116, 120
Crystal, Billy, 200
Curious Case of Benjamin Button, The, 54
Curtis, Tony, 222
Curtiz, Michael, 188

D

Damn Yankees, 80, 82
Dances with Wolves, 84, 150, 233-234
Dangerous Liaisons, 74
Dark Knight, The, 111-112, 130, 153, 240
Darkest Hour, 175
Davis Jr., Sammy, 198
Davis, Viola, x
Dawn of the Planet of the Apes, 11
Day-Lewis, Daniel, 173, 175, 178
De Niro, Robert, 230
Dead Poets Society, 173
Deakins, Roger, 73
Decline of the American Empire, 142
Deer Hunter, The, 149, 229, 243
Defiant Ones, The, 27
Del Toro, Guillermo, 239
Delmar, Vina, 167

DeMille, Cecil B., 63, 200-201
Demme, Jonathan, xii
Dench, Judi, ix, 28, 171, 175, 181
Departed, The, 75, 150, 152, 239, 249-250
DiCaprio, Leonardo, 202
Dickens, Charles, 166
Die Hard, 233
Directors Guild of America, 47-48, 189-190, 194-196, 199-203, 248-250
Dirty Dancing, 92
Disney, 66, 91, 122-123, 125-127, 235
Disney, Walt, 70
Django Unchained, 48
Do the Right Thing, 233, 244
Doctor Strange, 50
Dodsworth, 165
Dolby Theatre, 198, 202
Double Indemnity, 216, 218, 243, 247
Dove, The, 56
Dr. Strangelove, 180, 224, 242
Dreamgirls, 111-112
DreamWorks, 123, 126-127
Driving Miss Daisy, 2, 18, 150, 173, 200, 233
Duck Soup, 212-213, 244
Dukakis, Olympia, 152
Duke, Patty, 20
Dunaway, Faye, 5, 94, 197-198
Dunkirk, 112, 150
DuVernay, Ava, 90

E

East Lynne, 134
East of Eden, 28

Eastwood, Clint, 249
Ebert, Roger, 203
Ed Wood, 34
Education, An, 167
Edward, My Son, 74
Elephant Man, The, 32
Elizabeth, 35
Elmer Gantry, 165, 167
Empire Strikes Back, The, 230, 243
Enchanted, 67
English Patient, The, vii, 56, 150, 235
Enyedi, Ildiko, 143
Epstein, Julius, 216
Epstein, Philip, 216
Erin Brockovich, 179
Ernest & Celestine, 126
E.T. the Extra-Terrestrial, 231, 242
Eternal Sunshine of the Spotless Mind, 238
Evolution, 116
Ex Machina, 8
Executive Suite, 75
Exorcist, The, 59

F

Faith, Percy, 90
"Falling Slowly," 67
Fame, 99
"Fame," 98
Fantasia, 122
Fantastic Beasts and Where to Find Them, 49-50
Fantastic Woman, A, 142
Farewell My Concubine, 142
Farewell to Arms, A, 70
Fargo, 235-236, 243

Fatal Attraction, 74, 151
Fellini, Federico, 224
Fellowes, Julian, 24
Ferber, Edna, 165
Ferrer, Jose, 180
Ferris Bueller's Day Off, 232
Few Good Men, A, 59
Fiddler on the Roof, 79-80
Fight Club, 237, 244
Finch, Peter, 5, 171
Finding Nemo, 125
Finding Neverland, 85
Finney, Albert, 178-179
Firth, Colin, xi, 175, 240
Fish Called Wanda, A, 171
Five Easy Pieces, 226
Flashdance, 92
"Flashdance... What a Feeling," 92
Fleming, Victor, 65
Fletcher, Louise, xii
Flower Drum Song, 80
Folsey, George, 73, 75
Fonda, Henry, 18
Fonda, Jane, 171
Ford, Harrison, 230
Ford, John, 201, 215, 221
Forman, Milos, xii
Forrest Gump, 8, 53, 133, 150, 235, 243
Forster, E.M., 166
Fosse, Bob, 81, 227
Foster, Jodie, xii
Foster, Lewis R., 167
Four Weddings and a Funeral, 133
Fox, 204
Fox, Edward, 171

Freaks, 212
Free Soul, A, 74
Freeman, Morgan, 173
French Connection, The, 149, 162, 171, 226-227
Frida, 85
From Here to Eternity, 74, 148, 220
Front Line, 133
Front Page, The, 134
Frozen, 99, 126
Fun Home, 78
Funny Girl, 80
Funny Thing Happened on the Way to the Forum, A, 79-80

G

Gable, Clark, xii, 213
Gandhi, 33, 37, 149, 175, 178, 231
Gangs of New York, 191
Garland, Judy, 91-92, 214
Garson, Greer, 178
Gaslight, 138
Gay Divorcee, The, 98
Gaynor, Janet, 180, 204
General, The, 240, 243
Gentleman's Agreement, 135, 148, 167, 218
Gershwin, George, 202
Gershwin, Ira, 202
Ghost in the Shell, 122
Ghost World, 164
Giant, 119, 165
Gibson, Mel, 199
Gigi, 80-82, 135, 149, 191, 208, 222
Girl, Interrupted, 181
Gish, Lillian, 211

Gladiator, 8, 61-62, 111, 150, 237
"Glory," 90
Go-Between, The, 171
Godfather: Part II, The, 133, 149, 166, 209, 227-228, 242
Godfather, The (movie), 133, 149, 166, 195, 215, 227, 241
Godfather, The (book), 166
Going My Way, 148, 216-217
Gold Diggers of 1935, 98
Gold Rush, The, 241, 244
Golden Globes, vii-viii, ix, 130, 152, 190, 202-203, 248
Gone with the Wind, 9-10, 15, 24, 28-29, 106, 118-119, 148, 202, 214, 242
Goodbye, Mr. Chips (1939), 167
Goodbye, Mr. Chips (1969), 74
Goodfellas, 75, 233, 242
Gorgeous Hussy, The, 75
Gosford Park, 24-25
Gosling, Ryan, 90
Gossett Jr., Louis, 28
Graduate, The, 225, 243
Grahame, Gloria, 28
Grand Budapest Hotel, The, 54, 56, 85
Grand Hotel, 2, 148, 205, 212
Grapes of Wrath, The, 201, 215, 243
Gravity, 112-113, 150
Great Dictator, The, 215, 244
Great Expectations, 166
Great Gatsby, The, 64
Great Ziegfeld, The, 118, 148, 213
Greatest Show on Earth, The, 31, 60, 135, 148, 200-202, 219-220
Greed, 241, 245

Green Dolphin Street, 75
Green Mile, The, 166
Green Years, The, 75
Griffith, D.W., 240
Groundhog Day, 234, 243
Guardians of the Galaxy, 11
Guess Who's Coming to Dinner, 149, 197
Guys and Dolls (movie), 80
Guys and Dolls (musical), 93
Gwenn, Edmund, 19
Gypsy, 80

H

Hackman, Gene, 171
Hacksaw Ridge, 112
Haggis, Paul, 203
Hamlet, 148, 218
Hammerstein, Oscar, 81, 224
Hammett, Dashiell, 166-167
Handful of Dust, A, 171
Haneke, Michael, 143
Hanks, Tom, 235
Happy Feet, 127
"Happy Working Song," 67
Harden, Marcia Gay, 48, 181
Hardy, Tom, 27
Hart, Moss, 167
Harvey Girls, The, 91
Heartbeeps, 33
Heaven Can Wait (movie), 62, 68, 166
Heaven Can Wait (play), 166
Heaven Knows, Mr. Allison 74
Hellman, Lillian, 166
Hello, Dolly!, 79-80, 134
Help, The, x, 150, 152
Henry, Buck, 68

Henry, Justin, 229
Henry V, 173
Hepburn, Audrey, 224
Hepburn, Katharine, 18
Here Comes Mr. Jordan, 166
Herrman, Bernard, 222
"Hey Jude," 90
Hidden Figures, 50
"High Hopes," 88
High Noon, 201, 218, 220, 245
Hiller, Wendy, 27
Hilton, James, 167
History of Violence, A, 164
Hitchcock, Alfred, 57, 66, 74, 215, 220, 222-223
Hoch, Winton, 66
Hoffman, Dustin, 229
Hole in the Head, A, 88
Holiday Inn, 98
Homer, 162
Hope, Bob, 200
Hopkins, Anthony, xii
Hornby, Nick, 167
Hotel Terminus, 134
Hours, The, 181, 191
Houston, Whitney, 91
How Green Was My Valley, 148, 201, 205-207, 215, 239
How the Grinch Stole Christmas, 35
How the West Was Won, 110
How to Succeed in Business Without Really Trying, 93
How to Train Your Dragon, 127
Howard, Ron, 199
Howards End, 166, 175
Howl's Moving Castle, 126

Hud, 166-167
Hugo, 45, 112, 247-248
Human Comedy, The, 74
Hurt Locker, The, 104, 112, 240
Hurwitz, Justin, 90
Huston, John, 25, 201
Hutton, Timothy, 19
Hyer, Martha, 27

I

In Cold Blood, 167
In Old Arizona, 133-134
In the Bedroom, ix
In the Heat of the Night, 57-58, 225
Inarritu, Alejandro G., 143
Inception, 112, 115, 240
Incredibles, The, 46, 111, 125
Informer, The, 201
Inside Out, 125
Interstellar, 11
Intolerance, 240, 243
Iris, ix, 28
Irma la Douce, 80, 198
Iron Giant, The, 122
Iron Lady, The, x, 34, 70-71
It Happened One Night, xii, 60, 101, 148, 206, 213, 244
It's a Mad, Mad, Mad, Mad World, 110, 224
It's a Wonderful Life, 8, 217, 242
"(I've Had) The Time of My Life," 92
Ivory, James, 16

J

Jackson, Glenda, 171-172
Jackson, Peter, 191, 237
James Bond franchise, 89
Jannings, Emil, 180
Jaoui, Agnes, 143
Jaws, 228, 243
Jenkins, Barry, 158, 201
Jesus Christ Superstar, 80-81
Joan of Arc, 65-66
John, Elton, 91
Johnson, Ben, 171
Jolie, Angelina, 181
Judgment at Nuremberg, 162
Julia, 39, 166
Julius Caesar, 63
Jungle Book, The, 122
Juno, 150, 152
Jurassic Park, 12

K

Kaige, Chen, 142
Kaufman, Charlie, 179
Kazan, Elia, 220
Keaton, Buster, 240-241
Keaton, Michael, 171
Kelly, Gene, 86, 219
Kerr, Deborah, 74, 220
Kidman, Nicole, ix, 181
King and I, The, 74, 79-80
King Kong (1933), 110, 212, 244
King Kong (2005), 112-113
King Solomon's Mines, 70
King, Stephen, 166
King's Speech, The, xi, 150, 152, 175, 240, 253
Kingsley, Ben, 175, 178
Kiss Me Kate, 80
Klute, 171

Koch, Howard, 216
Kramer vs. Kramer, 70, 149, 229
Krizan, Kim, 167
Kubrick, Stanley, 222

L

La La Land, 38-39, 50, 62, 85, 90, 112, 150, 198-199, 201
"La Marseillaise," 89
Lady Be Good, 89
Lady for a Day, 198
Lady of the Tropics, 75
Laika, 123
Lancaster, Burt, 220
Lang, Charles, 70-71
Last Emperor, The, 84, 151, 191, 232
Last Picture Show, The, 166-167, 171
"Last Time I Saw Paris, The," 88-89
Lawrence of Arabia, 74-75, 118, 149, 169, 192, 223, 242
Leachman, Cloris, 171
Leave Her to Heaven, 70
Lee, Ang, 142-143, 203
Lee, Spike, 233
Legend, John, 90
Leighton, Margaret, 171
Lelio, Sebastian, 142
Lemmon, Jack, 222
Les Armateurs, 123, 126
Les Miserables, 112
"Let It Go," 99
Letters from Iwo Jima, 112, 249-250
Lewis, Sinclair, 165
Life of Emile Zola, The, 148, 162, 205-206, 214
Life of Pi, 195

Lilies of the Field, 224
Lincoln, 53, 59, 150, 177
Linklater, Richard, 166-167
Lion in Winter, The, 74, 162, 195
Lion King, The, 91, 122, 235, 245
Little Dorrit, 166
Little Foxes, The, 32, 166
Little Mermaid, The, 67, 92, 122
Little Miss Sunshine, 19, 249-250
Little Night Music, A, 79-80
"Livin' on a Prayer," 91
Llosa, Claudia, 143
Lloyd, Frank, 198
Lloyd, Harold, 110
Loesser, Frank, 93-94
Logan, 164
Logan's Run, 61-62
Lohan, Lindsay, 205
Lonely Passion of Judith Hearne, The, 171
Lonelyhearts, 27
Lord of the Rings: The Fellowship of the Ring, The, 85, 237-238, 244
Lord of the Rings: The Return of the King, The, 8, 42, 57, 85, 102, 119, 150, 152, 190-192, 238
Lord of the Rings: The Two Towers, The, 42, 191, 238
Lord of the Rings trilogy, *The*, 54
"Lose Yourself," 99
Lost in Translation, 191
Lost Weekend, The, 134, 138, 148, 162, 217
Love Finds Andy Hardy, 214
"Love Story," 91
Lucas, George, 229
"Lullaby of Broadway," 98

M

Maccari, Ruggero, 167
Mad Max: Fury Road, 56, 112
Madame Curie, 178
Madness of King George, The, 56
Magic Light, 126
Magnificent Ambersons, The, 216, 225, 245
Make-Up Artists and Hair Stylists Guild, 42, 45, 47-48
Malick, Terrence, 227
Maltese Falcon, The, 166-167, 215, 218, 243
Man for All Seasons, A, 149, 162, 225
Man of La Mancha, 80, 82
Man on Wire, 134
Mankiewicz, Joseph L., 201
Mann, Abby, 162, 168
Marjoe, 134
Martian, The, 63
Marty, 3, 118-119, 134, 151, 221
Marvel, 50
Marvin, Lee, 180
Marx Brothers, 212-213
Mary Poppins, 55, 80, 89, 224
Master and Commander: The Far Side of the World, 42, 191
Mating Season, The, 74
Matlin, Marlee, 18
Matrix, The, 5, 237, 243
McCambridge, Mercedes, 28
McCraney, Tarell Alvin, 158
McDaniel, Hattie, 24, 28-29
McDormand, Frances, 19
McMurtry, Larry, 166-167
McQueen, Steve, 94

Meet Me in St. Louis, 75, 80, 167
Memento, 5, 237, 244
Menken, Alan, 66-67, 92
Mercer, Johnny, 92
Metro-Goldwyn-Mayer, 212
Midnight Cowboy, 134, 149, 226
Midnight in Paris, 51
Milagro Beanfield War, The, 85
Mildred Pierce, 217
Milk of Sorrow, The, 143
Milland, Ray, 217
Million Dollar Baby, 59, 150, 238
Million Dollar Mermaid, 75
Mills, John, 28
Miracle on 34th Street, 19, 145
Miracle Worker, The, 20
Mirren, Helen, 24-25, 175
Mirror Has Two Faces, The, vii
Mister Roberts, 151
Miyazaki, Hayao, 126
Modern Times, 110, 213, 243
Moneyball, viii
Monster's Ball, ix
Monsters Inc., 126
Moonlight, 24, 28, 60, 158, 198-199, 201
Moonstruck, 149, 151-152
Moore, Julianne, 18, 175
More the Merrier, The, 167
Motion Picture Sound Editors, 46-49, 112-113
Moulin Rouge, 180, 201
Moulin Rouge!, ix
Mr. Smith Goes to Washington, 59, 167, 214, 217, 245
Mrs. Doubtfire, 37

Mrs. Miniver, 148, 167, 216
Music Man, The, 79-80
Mutiny on the Bounty, 148, 213
My Fair Lady, 79-80, 109, 149, 166, 224
My Family, 35-36
My Favorite Year, 74
"My Heart Will Go On," 93
My Left Foot, 173, 175
Mystic River, 191

N
Nashville, 228, 244
National Board of Review, 205-206
National Velvet, 74
Natural, The, 53, 74
Neptune's Daughter, 93-94
Network, 1, 5, 197
New York Film Critics Circle, 205
Nicholas and Alexandra, 56
Nicholson, Jack, xii
Nickelodeon, 123, 126
Night at the Opera, A, 213
Night of the Hunter, The, 221, 244
No Country for Old Men, 69, 112
Nolan, Christopher, 5, 130
Nolan, Jonathan, 5
North by Northwest, 222-223, 242
Nyong'o, Lupita, 179

O
O Brother, Where Art Thou?, 162, 164
Odd Couple, The, 166
Odyssey, The, 162
Officer and a Gentleman, An, 28
O'Hara, Maureen, 201

O.J.: Made in America, 133-134
Oldman, Gary, 175
Oliver!, 80, 135, 149, 164, 166, 192, 195, 225-226
Olivier, Laurence, 218
On Body and Soul, 142-143
On Golden Pond, 18
"On the Atchison, Topeka, and the Santa Fe," 91
On the Waterfront, 148, 220, 242
Once, 67
One Flew Over the Cuckoo's Nest, xii, 149, 213, 228, 242
One Hundred Men and a Girl, 78
"One Moment in Time," 91
One Night of Love, 78
O'Neal, Ryan, 21
O'Neal, Tatum, 20-21
Only Yesterday, 122
Operator 13, 75
Ordinary People, 19, 25, 59, 149, 230
Osment, Haley Joel, 205
Ostlund, Ruben, 143
O'Toole, Peter, 74-75
Out of Africa, 84, 149, 180, 195, 232
Out of the Past, 218, 245
"Over the Rainbow," 98

P
Palance, Jack, 28
Palin, Michael, 171
Pan's Labyrinth, 239
Paper Moon, 20
Paquin, Anna, 20
Parent Trap, The, 179, 205
Pasek, Benj, 90

Passage to India, A, 20, 166, 175
Paths of Glory, 222, 244
Patriot, The, 134
Patton, 149, 226
Paul, Justin, 90
Persepolis, 127
Philadelphia, 37
Phoenix, Joaquin, 152
Pianist, The, 17, 181, 191-192, 195
Piano, The, 20
Pickup on South Street, 74
Picnic, 148, 151
Pillow Talk, 74
Pinocchio, 78, 122
Pixar, 121, 123, 125-128
Place in the Sun, A, 202, 219
Platoon, 110, 149, 200, 232
Pleasantville, 42
Plummer, Christopher, 16, 19
Pocahontas, 67, 91
Pollock, 48, 181
Price, Richard, 167
Prime of Miss Jean Brodie, The, 175
Prince, 78
Prince of Egypt, The, 67
Princess Bride, The, 232-233
Producers Guild of America, 43, 47-48, 201, 248-249
Professionals, The, 167
Profumo di donna, 167
Psycho, 215, 223, 242
Pulp Fiction, 133, 235, 242
Purple Rain, 78
Puzo, Mario, 166-167
Pygmalion (movie), 166
Pygmalion (play), 166

Q

Queen, The, 175, 249-250
Quest for Fire, 33, 37
Quiet Man, The, 66, 200-201
Quiz Show, 133

R

Rademakers, Fons, 142
Raging Bull, 75, 230, 242
Raiders of the Lost Ark, 230, 243
Rain Man, 150, 233
Random Harvest, 167
Rango, 126
Ratatouille, 127
Reap the Wild Wind, 66
Rear Window, 215, 220, 242
Rebecca, 57-58, 134, 148, 215
Redford, Robert, 25, 232
Redmayne, Eddie, 171, 175
Reds, 25, 28, 197
Reunion in Vienna, 75
Revenant, The, 27, 112-113, 150, 201-202
Rice, Tim, 91
Right Stuff, The, 231, 238
Risi, Dino, 167
Ritter, Thelma, 74
RKO, 216
Robbins, Jerome, 68
Rockwell, Sam, 19
Rocky, 149, 228, 245
Rodgers, Richard, 81, 224
Rogers, Will, 198
Romance, 74
Rome, Open City, 138
Room with a View, A, 166

Roommates, 35-36
Rooney, Mickey, 214
Rope, 66
'Round Midnight, 85
Roxanne, 157
Ruffalo, Mark, 27
Ruling Class, The, 74
Russell, Harold, 28-29
Ryan's Daughter, 28
Rylance, Mark, 27

S

Safety Last!, 110
Saint, Eva Marie, 220
Saving Private Ryan, 35, 110, 203, 236, 243
Scent of a Woman, 167
Schindler's List, 37, 56, 85, 150, 234, 242
Schwartz, Stephen, 66-67
Scorsese, Martin, 75, 228, 230, 239, 250
Scott, A.O., 177-178
Scott, Ridley, 230
Screen Actors Guild, vii-viii, ix, 43, 47-48, 200-201, 203, 249-250
Seabiscuit, 191
Searchers, The, 221, 242
Secret in Their Eyes, The, 143
Segall, Harry, 166
Selfish Giant, The, 116
Sellers, Peter, 180
Selma, 90
Separate Tables, 27, 74
Seven Brides for Seven Brothers, 75, 80
Seven Samurai, 221, 244
Shadow of the Vampire, 35

Shakespeare in Love, 28, 35, 42, 150, 175, 181, 203, 236
Shakespeare, William, 218, 236
Shamroy, Leon, 70-71
Shane, 220
Shani, Yaron, 143
Shape of Water, The, 57, 189, 200, 239
Shark Tale, 124-125
Shaw, George Bernard, 166
Shawshank Redemption, The, 133, 166, 235, 242
She Done Him Wrong, 3, 119
She Wore a Yellow Ribbon, 66
Sherlock Jr., 241, 244
Shining, The, 230, 244
Shootist, The, vii
Shrek, 126, 164
Silence of the Lambs, The, xii, 150, 213, 234, 243
Simon, Neil, 166
Sinatra, Frank, 86, 88
Singin' in the Rain, 88, 219, 242
Sixth Sense, The, 205, 237
Skall, William, 66
Skippy, 134, 164
Skyfall, 46, 89, 112
"Skyfall," 89
Slumdog Millionaire, 85, 112, 119, 150, 152-153, 240
Smith, Maggie, 171, 175
Snow White and the Seven Dwarfs, 122, 214, 235
"So Close," 67
Social Network, The, 85
Some Came Running, 27
Some Like It Hot, 222-223, 242

Somebody Up There Likes Me, 56
Sondergaard, Gale, 24, 26
Song of Bernadette, The, 39
Song of the Sea, 126
Sony, 123, 127
Sophie's Choice, 70
Sound of Music, The, vii, 2, 10, 79-80, 107, 110, 149, 224, 243
South Pacific, 80-82
Spacek, Sissy, ix
Spacey, Kevin, 16
Spartacus, 65
Spielberg, Steven, 234
Spirited Away, 124-125
Spotlight, 27, 201-202
Square, The, 143
Stage Door, 165
Stallone, Sylvester, 27
Stand by Me, 166
Stapleton, Maureen, 27-28
Star Trek, 61-62
Star Wars, 10, 39, 56, 229, 242
Star Wars: The Force Awakens, 9
Star Wars: The Last Jedi, 103
State Fair, 80
Steinbeck, John, 215
Stevens, George, 202
Stewart, Jimmy, 217, 222
Still Alice, 18, 175
Sting, The, 25, 59, 149, 227
Stone, Oliver, 200
Stop the World – I Want to Get Off, 80
Story of Louis Pasteur, The, 162
Streep, Meryl, x-xi, 70-71, 229, 232
Street Angel, 204
Streetcar Named Desire, A, 166, 219

Studio Ghibli, 123, 125-126
Stunt Man, The, 74
Sullivan's Travels, 215, 244
Sunday Bloody Sunday, 171
Sundowners, The, 74
Sunrise, 110, 204, 210-211, 243
Sunset Boulevard, 155, 219, 242
Sunshine Boys, The, 166, 181
Surf's Up, 127
Surtees, Robert, 70
Swank, Hilary, 177-178
Sweet Charity, 80
Swift, Taylor, 91
Syriana, 25

T

Tandy, Jessica, 18, 233
Tarantino, Quentin, 235
Taste of Others, The, 143
Taxi Driver, 75, 228, 242
Taxi to the Dark Side, 134
Tempest, 56
Ten Commandments, The, 62, 119
Terms of Endearment, 149, 166-167, 231
"That's How You Know," 67
"Theme from *A Summer Place*," 90
Theory of Everything, The, 171, 175
Thin Man, The, 166-167
Third Man, The, 161, 225
Thomas Crown Affair, The, 94
Thompson, Emma, 175
Thoroughly Modern Millie, 80
Thousands Cheer, 75
Three Billboards Outside Ebbing, Missouri, 19
Tin Pan Alley, 78

Titanic, 2, 8, 11, 34, 93, 101-102, 104, 150, 236, 244
To Have and Have Not, vii
To Kill a Mockingbird, 223, 243
To the Shores of Tripoli, 66
Tolkien, J.R.R., 54
Tom Jones, 120, 149, 198, 224
Tommy, 80
Touch of Class, A, 172
Touch of Evil, 208, 222, 225, 244
Toy Story, 122, 235, 243
Toy Story 2, 122
Toy Story 3, 58, 122, 127, 240
Trader Horn, 134
Traffic, 192, 195
Treasure of the Sierra Madre, The, 25, 47, 218, 243
Trouble in Paradise, 212, 244
True Grit (1969), 18
True Grit (2010), 69, 112
Trueba, Fernando, 142
Turning Point, The, 39

U

"Under the Sea," 67, 92
Unforgiven, 150, 234, 244
Up, 58, 85, 110, 122, 125, 240
Usual Suspects, The, 158, 160, 187, 235, 245

V

Valentine, Joseph, 66
Van Fleet, Jo, 28
Van Heusen, Jimmy, 88
Venus, 74
Vertigo, 208, 215, 222, 241

Vidor, King, 74, 205, 211
Visual Effects Society, 45, 47-48
Von Stroheim, Erich, 241

W

Walk the Line, 18, 112, 150, 152
Wallace and Gromit, 126
WALL-E, 125
Waltz, Christoph, 48
War Horse, 112
War of the Worlds, 112
Warner Brothers, 98, 224
Warren, Harry, 91-92
Watch on the Rhine, 166-167
Wayne, John, vii, 18, 201, 221
Webber, Andrew Lloyd, 81
Wedding Banquet, The, 142
Weir, Andy, 63
Welles, Orson, 188, 205, 215-216, 222, 225
West Side Story, 5, 19, 68, 80, 149, 181, 223, 244
"When You Believe," 67
Whiplash, 111-112
"White Christmas," 98
White Cliffs of Dover, The, 75
White Heat, 218
White Ribbon, The, 143
Who Framed Roger Rabbit, 55, 121-122
"Whole New World, A," 67, 99
Who's Afraid of Virginia Woolf?, 225
Wild Bunch, The, 226, 244
Wilder, Billy, 222-223
Williams, Cara, 27
Williams, John, 70

Williams, Robin, 37, 173
Williams, Tennessee, 166
Willis, Bruce, 233
Wilson, 70
Wind, The, 211
"Windmills of Your Mind," 94, 96-97
Wings, 2, 8-9, 57, 148, 204, 210-211
Wise, Robert, 68
With a Song in My Heart, 74
Witherspoon, Reese, 18, 152
Wiz, The, 80
Wizard of Oz, The, 41, 78, 98, 138, 214, 242
Wolf of Wall Street, The, 119
Woman Under the Influence, A, 228, 244
Women in Love, 172
Woodstock, 134
World According to Garp, The, 74
Wreck-It Ralph, 127
Writers Guild of America, 43, 47-48, 155-160, 199

X

X-Men: Days of Future Past, 11

Y

Yankee Doodle Dandy, 23
Yearling, The, 74
Yellow Submarine, 122
You Can't Take It with You, 148, 167, 214
You Light Up My Life, 89
"You Light Up My Life," 89

Z

Z, 134
Zero Dark Thirty, 46, 112, 179
Zeta-Jones, Catherine, 181
Zinneman, Fred, 201
Zootopia, 125

Printed in Poland
by Amazon Fulfillment
Poland Sp. z o.o., Wrocław